AF559870

BASIC CHEMISTRY

BASIC CHEMISTRY

By

Shardendu Kislaya

DISCOVERY PUBLISHING HOUSE PVT. LTD.
NEW DELHI-110 002

Published by:
Tilak Wasan
DISCOVERY PUBLISHING HOUSE PVT. LTD.
4383/4B, Ansari Road, Darya Ganj
New Delhi-110 002 (India)
Phone : +91-11-23279245, 23253475, 43596065
E-mail : discoverybooksindia@gmail.com
discoverypublishinghouse@gmail.com
web : www.discoverypublishinggroup.com

***Reprinted:* 2019**

***First Published:* 2011**

ISBN: 978-81-8356-753-4

Basic Chemistry

Printed at:
Infinity Imaging Systems
Delhi

Preface

In many high schools and colleges the basic chemistry course is the one that causes most concern among students. With everything going right, chemistry can be a fun but challenging course. Under poor conditions, your first chemistry course can be a real spittin, cussin nightmare. The study of chemistry may be different from anything else you have ever done. In fact, there are several different topics in chemistry that may need types of studying tailored to the topic. Basic chemistry is a survey course in chemistry (That is, it covers a little bit of everything.) with emphasis on common chemicals and study techniques. The aim is to give you some chemical and general scientific literacy rather than train you to be a chemist. This outline is to:

(a) tell you what to expect in most courses;

(b) show you some methods of study for the course; and

(c) show you some directions you can turn to for help, if you should need it.

The most useful advice is to stay up with or ahead of the class. As the Red Queen said to Alice, "Now, here, you see, it takes all the running you can do, to keep in the same place. If you want to get somewhere else, you must run at least twice as fast as that!" If you fall significantly behind in basic chemistry, it is hard to catch up, but catching up is the only way to continue. The material "snowballs" in this course. By that I mean that the basic facts that need to be learned or memorized at the beginning are going to be used later in the course. Fluency in the basic material is necessary for you to be able to understand and learn the more complex ideas in the course. Chemtutor can help you keep up by showing you study methods, explanations, and other ways to learn the material.

We have found that students do better by having a quiz over a small amount of material. Act as if you will have a quiz every class period. These quizzes would either ask you to memorize some basic material or to use some of the material in a math process or a basic idea about chemistry. Designing and giving yourself quizzes to help you schedule your studying will prepare you for the major tests. Chemtutor usually has a Quickquiz or some math problems available to you to help you. If you are studying with someone, teach it to each other. The teacher always learns more than the student. (If you have it together sufficiently well to present it to someone else, you know it a lot better.)

Many students find studying with others from the same class is a lot of help. If you get together in a group of three or four, you will always have someone to study with. If you study with other students, it is easier to call them for missed assignments if you must be out.

Author

Contents

CHAPTER 1

Introduction

In many high schools and colleges the basic chemistry course is the one that causes most concern among students. With everything going right, chemistry can be a fun but challenging course. Under poor conditions, your first chemistry course can be a real spittin, cussin nightmare. The study of chemistry may be different from anything else you have ever done. In fact, there are several different topics in chemistry that may need types of studying tailored to the topic. Basic chemistry is a survey course in chemistry (That is, it covers a little bit of everything.) with emphasis on common chemicals and study techniques. The aim is to give you some chemical and general scientific literacy rather than train you to be a chemist. This outline is to:

(a) tell you what to expect in most courses;

(b) show you some methods of study for the course; and

(c) show you some directions you can turn to for help, if you should need it.

The most useful advice is to stay up with or ahead of the class. As the Red Queen said to Alice, "Now, here, you see, it takes all the running you can do, to keep in the same place. If you want to get somewhere else, you must run at least twice as fast as that!" If you fall significantly behind in basic chemistry, it is hard to catch up, but catching up is the only way to continue. The material "snowballs" in this course. By that it means that the basic facts that need to be learned or memorized at the beginning are going to be used later in the course. Fluency in the basic material is necessary for you to be able to understand and learn the more complex ideas in the course. Chemtutor can help you keep up by showing you study methods, explanations, and other ways to learn the material.

We have found that students do better by having a quiz over a small amount of material. Act as if you will have a quiz every class period. These quizzes would either ask you to memorize some basic material or to use some of the material in a math process or a basic idea about chemistry. Designing and giving yourself quizzes to help you schedule your studying will prepare you for the major tests. Chemtutor usually has a Quickquiz or some math problems available to you to help you. If you are studying with someone, teach it to each other. The teacher always learns more than the student. (If you have it together sufficiently well to present it to someone else, you know it a lot better.)

Many students find studying with others from the same class is a lot of help. If you get together in a group of three or four, you will always have someone to study with. If you study with other students, it is easier to call them for missed assignments if you must be out.

Learning Style Management

Learning is easy for some people. They can read something or hear something and understand it. For others the number of useful learning methods are more limited. Just because a person has a limited way to easily learn does not mean the person is stupied or uneducable. The term LD (learning disability) indicates that a person has a difficult time acquiring and learning information by one source or another.

It is truly unfortunate that a small number of students have used the idea of a learning disorder to attempt to cover for laziness. Most of the students who have a learning disorder would dearly love to be like other students, but they have experienced disappointment after disappointment. If you have a learning disorder, you must analyze (or have analyzed) exactly what it is and learn your best way to compensate for it. Anyone teaching for any time has seen the student who will obey a spoken request, but when asked to read the request will appear to let it go in one eye and out the other. For some people the written word just does not register in the right way. For this type of learning disorder a student might find that reading the assignment aloud from the book will help. For others, copying the words in the book works best.

For students who have a rough time understanding oral lectures, bring a tape recorder to school and tape the lecture. At home, speak the lecture after the teacher or transcribe the lecture. Whatever works is best for you. Particularly the students who want to go into any type of scientific work will see this type of material again along with the need to grasp great amounts of it quickly. So the basic chemistry course makes a good practice ground for whether you can adapt yourself to this type of study. You know yourself best, so the person to help you with learning problems best is yourself. You need to vary your learning techniques until you find something that works for you for each type of studying. Try a number of methods. Use varying techniques. If teaching chemistry to your little sister works for you, do it. If you do not seem to find a way that you can use, consult a counselor who is willing to help you get professional assistance.

How to Study for Tests

The type of test you can expect will, to a certain extent, determine what kind of studying you must do for it. Most instructors will tell you what style of test they will give you.

Objective tests (fill-in-the-blank or multiple choice) likely have the most obvious study methods. Flash card and association techniques are old standards for this type of work. Students working in pairs can quiz each other on this type of material. Make lists when you need to.

Be sure you understand definitions rather than just memorizing them. Math tests require the information of the objective tests and the practice in working problems. Historically in this class, the majority of trouble with math comes not from the math itself, but with background information that is lacking. Much of the background comes from the material Chemtutor suggests you know. If you have a problem with the math, first go back and learn the background well enough to use it. If the math itself is a problem, you need to ask for help from a student or instructor.

There is some help in math for you in Chemtutor, but if you have live help who can emphasize the points your instructor is requiring, that can be as useful. Many times your instructor will suggest a problem solving technique to you. Learn it and use it. If your instructor does not suggest a problem solving technique, Chemtutor has several approaches you can use.

Essay questions require yet another set of skills. The questions contain words like "explain" or "describe" or "compare", and you will be expected to write an essay on the subject at the test. The first requirement, of course, is that you be able to write in the English language. Without intending to be unfriendly or mean. Some students are used to writing a large amount of FILLER on essay tests.

Most instructors can easily see through that and do not grade it, except downward if it becomes excessive. Most instructors look for content when they grade an essay test. Essay answers may be expected to contain information from the book and/or from lecture. To study for essay tests, you should:

(a) consider some of the likely questions you might have;
(b) collect information;
(c) organize the information into an answer;
(d) compare your answer to some of the answers of your classmates; and
(e) practice answering the questions at home as if you were at the test.

You should memorize important items you need to mention ("key words"), but usually efforts to memorize whole answers word-for-word do not work well.

Chemtutor can help you keep up with the class. If you still need help, other students in your class may be able to help you. If you do not get the help you need from students, call or visit your instructor. If the other students can not help you, maybe they also need some help. Most instructors would be glad to help you on an individual or group basis. Get help far before you find yourself close to failing. Many instructors understandably feel uncomfortable about helping students with their work only a few hours or minutes before a test. Usually they havetest materials to get together at that time and feel a panicky student is not likely to learn much at short notice.

Class Management

Chemtutor can be of significant help to you in the general material of the course, but each course and instructor is somewhat different. Your instructor should give you a course syllabus, a description of the course with information on what you need to do to pass the course. Keep your course syllabus as the first page of your class notebook. This may sound somewhat compulsive, but a well-kept notebook can be a great help to you. Keep all your notes in a three-ring or other insertable binder. Everything pertaining to the course should go into the notebook, preferably in date order.

Taking class notes is another important art. You must be able to judge what is important enough to write in your notes. If you are too busy writing notes, you may miss something useful. If you have trouble taking notes, you should compare your notes with the notes of other students in the same class as soon after the class as you can.

If you need help or information (notes or assignments) in the course, your first questions should go to the members of your lab or study group. Next, ask other students in the class. Get the name and phone number or email address of the members of your lab group and other students in the class. Keep these in your notebook.

Some colleges and universities allow you to drop a course before the end of it to preserve your grade average. If for some reason you need to drop out of the course, give your instructor the courtesy of the opportunity to talk with you. There are some good reasons for wanting to drop out; you don't feel you have the background, you don't have the time to devote to the work, you have other turmoil in your life and can't expend the emotional or mental effort, your work schedule conflicts with your school, etc.

Somc Pointers on General Studying for Chemistry

- Keep up with the course. Catch up as quickly as you can if you fall behind.
- Know all rote material perfectly.
- Understand all concepts before going on.
- Thoroughly memorize all background material you are assigned.You will need it later.
- Keep a close watch on the assignments for the course. Call other students if you must.
- Find a good place to study. It must be comfortable, quiet, well-lit, and have all the things you need.

- Find a good consistent time for studying chemistry.
- Review lectures as soon afterward as you can.
- Determine which study methods are best for you.
- Study with others if it helps you. Choose study partners who are serious about learning.

Problem Solving Skills in Chemistry

In general, accuracy needs to increase rather than speed. Any technical literature is concentrated, and any form of "speed reading" is a waste of time on informationally dense material. Instead, devices to increase comprehension are necessary, such as mental models, finger counting, lip movement, figure drawing, or reading aloud to another person.

Intelligence may be defined as the ability to do abstract reasoning. Weaker students jumble abstract reasoning from lack of ability to grasp the entire problem at "one shot". The next step for a weaker student is to give up for lack of a way to even get started. The best students support abstraction with concrete ideas. It is not the ability of the better student to fully grasp the complete abstraction that sets them apart, but the ability to organise a problem so that no part of it is too difficult. The idea behind the abstraction becomes more apparent as the idea is used. Another way to put that is that the better student does not usually have more efficient mental "hardware" but better "software".

One tool for the development of better methods of problem solving is to take a short standard intelligence test and later analyze the results. The number of correct answers is not the most important data, but the analysis of how to "concrete" each question and how to spot potential errors on a case-by-case basis can help students see some ways to improve their personal analytical reasoning ability.

The use of a standardized test will give an ability to see a broad general group of problems that are each "informationally neutral", that is, having no need for a body of background information other than the very rudimentary, such as the number system, the alphabet, and the ability to define words. Indeed, on any IQ test some questions require just word definition knowledge. In coursework, though, a body of knowledge is needed to interpret the questions. In chemistry some of this material may have to be known by rote in order to most efficiently perform on tests. Some examples of material that should be know by rote are the symbols of elements, polyatomic ions, some valences, measurements and the conversion factors among them, dimensions and the symbols for them, and some common names for materials. There are several ways to learn rote material, to include flashcards, pair quizzing, mnemonic devices, and reading aloud. Class time is poorly spent on rote material, but it is the teacher's responsibility to point out which material is a candidate for rote learning.

There is, unfortunately, no way to "pour" this information into a student. That basic maturity of recognizing something that must be done and doing it is necessary as a prerequisite. We say that the math is difficult for the students, but they can do the arithmetic very well on the calculators they have. The real stumbling blocks are difficulties organizing the problem and a lack of background rote information. The problem-solving technique needs to be practiced first with trivial problems and then with increasingly difficult problems. "Practice makes perfect" seems to be true; the best way to learn what can go wrong in a problem is to make the mistake yourself, find the mistake, and learn from your mistake.

Again, from the initial observations, any problem that can not be thought out completely in the head needs an overall 'roadmap' toward a solution and an orderly implementation of the pathway in which each step is demonstrable. For some good ideas on solving problems that are not the "formal" type of chemistry homework problem, see Polya's book, "How to Solve It." For chemistry problems involving a formula, one pathway is the W5P method to be introduced later. For most conversion problems (many of the chemistry problems), the Dimensional Analysis system, also to be introduced later, is a splendid framework. The point is that with a good framework in which to think of the problem, a complicated problem is merely a series of simple problems.

Some possibly useful pointers on problem solving:

- Read the problem carefully, moving your lips if necessary.
- Find the best way to think of the problem in the most concrete terms.
- Draw the problem, if possible.
- List the important information items in the problem.
- Know the background material for the problems.
- Follow the problem-solving method your teacher gives you, if one is given.
- Understand and use the units of quantities.
- Know your own weakest points so you can be wary of them.
- Go back and check your work. Use the units of the measurements to check your work. If the answer does not make sense, something is wrong.

The universe is composed of essentially three things: matter, energy, and empty space. Chemistry is the study of how matter interacts and we need to understand some of the basic rules and ideas about matter to understand how living things work. An element is the most basic form of matter. It is a substance that cannot be separated into simpler substances by chemical means. A compound is a combination of two or more elements. A typical living thing is mainly composed (99.9% by weight) of just six elements: carbon, hydrogen, nitrogen, oxygen, phosphorus, and sulphur.

Seventeen other elements occur in minute quantities or as just traces. Elements are composed of identical particles are called atoms. We can define atoms as the smallest particle into which an element can be divided and still have the properties of that element. A molecule is two or more atoms joined together by something called molecular bonds.

Atomic Organization

Atoms are made up of smaller (or subatomic) particles and the three most important are called protons, neutrons, and electrons. A Proton - is a positively charged particle with a weight of one atomic mass unit. A Neutron - has no charge and a weight of one atomic mass unit (weighs the same as a proton). An electron - is a negatively charged particle with a weight 1/2000th of a proton (we will consider it to have no weight).

It is important to remember that particles with the same charge will repel each other and particles with opposite charges (like protons and electrons) will attract each other. Protons and the electrically neutral neutrons make up the center or nucleus of an atom and the electrons occur in orbits moving around the nucleus; they are held there by the attraction of oppositely charged particles. The Atomic Number of an element is the number of protons in the nucleus of its atoms. Since hydrogen, the simplest atom, has only a single proton in its nucleus, its atomic number is 1. Oxygen is 8 and carbon is 6. Note: in a "normal" atom the number of electrons equals the number of protons; therefore "normal" atoms have no net charge - the positive cancels the negative.

The Atomic Mass of an element is the sum of the number of protons and the number of neutrons, since each has a weight of 1 atomic mass unit (amu).

Atomic Variants

1. *Isotopes* - isotopes of an element differ in the number of neutrons in the nucleus. For example, uranium

normally exists as U*238* with 92 protons and 146 neutrons. An important isotope of U*238* is U*235* with three fewer neutons. This isotope is radioactive; its nucleus is unstable and emits energy (which humans have used for power plants and bombs).

2. *Ions* - ions are atomic variants with either a net negative or a net positive charge because the number of electrons differs from the number of protons. Sodium (Na) normally has 11 protons and 11 electrons. One biologically important ion is symbolzed as Na^+; it has lost an electron and so has a net positive charge.

Electrons in Orbit

What gives each atom its unique chemical behaviour? Why is hydrogen an odorless gas and carbon a black powdery substance? The answer lies in the number and position of electrons in the energy levels (or shells), especially in the outer energy level, the one in closest contact with the rest of the universe.

Electrons swirl around an atom's nucleus in specific paths called orbitals (orbitals are where the electron spends most of its time). The innermost energy level can only hold two electrons in one orbital, the next energy level can hold eight in 4 different orbitals, the third can hold eight, and so on. When these energy levels contain their maximum number of electrons, they are said to be filled. If the outermost energy level of any particular atom is filled, that element is said to be inert meaning it is very stable and generally can only interact with other atoms with a great deal of added energy. Hydrogen has only one electron, so its outermost energy level is not filled; it can hold one more. Consequently, hydrogen can react with other atoms to form molecules. Helium has two electrons in its one energy level, therefore it is filled; helium is inert. Other inert elements are neon and argon, sometimes called the Noble gases because they do not react with other elements easily; they are "nobility". Carbon has two electrons in its innermost level and four in its outermost level for a total of six. Therefore, its outermost level needs four more electrons to be filled; carbon can combine (share electrons) with up to four more atoms; it can form a huge number of different compounds. Sodium has only one electron in its outer level (can hold eight); it is very unstable. Chlorine has seven electrons in its outer energy level; it needs only one to fill it. As you might guess, sodium and chlorine react very easily with each other.

Most atoms with unfilled energy levels tend to react (we will discuss chemical reactions later) in such a way as to completely fill the outer energy level when the reaction is over. We will discuss energy later, but for now understand that the further an electron is from the nucleus, the more potential energy it possess. To understand this concept, it may be useful to make an analogy to a boulder that is pushed up to a hill top. When it arrives there, the boulder has a certain amount of potential energy. If it is allowed to roll down the hill, the potential energy is converted to kinetic energy (i.e can do work, like smash things at the bottom of the hill). Just like the rock that will tend to move to the bottom of the hill, electrons tend to seek their lowest unfilled energy level, the one closest to the nucleus. It takes energy to move them to higher energy levels; unlike the rock, an electron cannot go partway up a hill; it must go to a discrete energy level.

This discrete amount of energy needed to move an electron from one level to the next highest is called a quantum. Later, we will see how the energy in sunlight can be used by green plants to raise electrons to higher energy levels - this is the source of almost all chemical energy available to living things.

Molecular Bonds

Molecules are formed when two or more atoms are bonded together. These bonds are energy links between the atoms, not physical couplings (like glue). There are three main types of chemical bonds.

1. *Covalent Bonds:* the most common type of bond. They occur when a pair of electrons is shared by two or more atoms. A single bond, occurrs when one electron is shared between, for example, a hydrogen and the carbon atom. Double bonds occur when two electrons are shared between a pair of atoms.

   If the sharing is equal between the atoms, we say the bond (and the molecule) is nonpolar. The electrical charge (the electron) is equally shared with both ends (the poles)of the molecule. If the shared electrons spend more time orbiting one atom in the molecule over another, the bond (and the molecule) is said to be polar. For example, the electrons shared between oxygen and hydrogen in a water molecule spend more time with the oxygen atom than the hydrogen atom; the oxygen pole of the molecule has a slight negative charge (because the electron has a negative charge) while the hydrogen pole has a slight positive charge (because at times it has no electron, just a positively-charged proton).

2. The polarity mentioned above leads to another kind of bond called a hydrogen bond. This is a bond between the negative pole of a polar molecule and the slight positive charge on a hydrogen atom that is participating in another polar molecule. For example, the oxygen in a water molecule (which is polar) will form a hydrogen bond with a hydrogen atom of another water molecule. These bonds are weak and easily broken. In any given quantity of liquid water, countless hydrogen bonds are forming and breaking all the time. These bonds give water many unusual properties.

3. Ionic bonds are the third type of chemical bond important for our understanding of chemistry. Here there is a complete transfer of electrons from one

molecule to another. It is most common when one molecule needs just one or two to complete its outermost shell and the other has only one or two in its outermost shell, like the situation in chlorine and sodium described earlier. When sodium come in contact with chlorine, the sodium loses one electron and the chlorine gains one. Now the two atoms are oppositely-charged ions so they will remain in contact (like magnets) to form the compound sodium chloride (NaCl or table salt). Ionic bonds are not as strong as covalent bonds in many situations, for example they will dissociate (fall apart) when dissolved in water.

Before we begin a discussion of organic compounds (compounds containing carbon) that constitute living things, we need to briefly discuss one of the most important compounds for life: water.

Water Chemistry

Life began in water and organisms are composed of anywhere from 50 to 90% water. Today, so many kinds of living things live in water. What is so special about it and why is it so special chemically? The answer lies in the hydrogen bonding between water molecules and other water molecules as well as with other substances; recall water is a polar compound, it has positive and negative ends.

Properties of Water

1. Water is slow to heat up and slow to release heat compared to other liquids. More heat is needed to raise the temperature of a quantity of water than other liquids. Most of the heat energy applied to water is used to break the hydrogen bonds before the temperature can be raised. Water is said to have among the highest specific heat of compounds that normally exist as a liquid. This ability of water is important to organisms because it keeps the environment in and around them more constant. Rapid temperature fluctuations impede many

life processes; few organisms can cope with the physiological stresses associated with them.

2. Water has a high heat of vaporization. This means that when water evaporates, a great deal of heat is required; this heat is drawn from the surrounding environment. Organisms can use this property to cool themselves (sweating) by allowing water to evaporate from their bodies.

3. Water molecules exhibit a high degree of cohesion (tendency of water molecules to cling to each other) and adhesion (tendency of unlike molecules to cling to each other). These properties account for capillarity, where water moves upward against the pull of gravity in narrow spaces. Plants use this property of water to move materials up their stems from the roots to the leaves.

4. Water exhibits a high surface tension - the tendency of molecules at the surface of a liquid to cohere to each other and not to the air above. Some organisms can utilize this property and literally walk on water!

5. Ice floats! Ice is less dense than liquid water and hence floats in it. If it didn't, lakes would freeze solid each winter, resulting in grave consequences for organisms that live in lakes.

6. Water is sometimes referred to as the universal solvent because so many substances (both ionic molecules and many polar covalent molecules) can dissolve in it. A solvent is a substance capable of dissolving other substances while the solute is the substance that is dissolved. Together they make a solution. Hydrophilic ("water-loving") substances readily dissolve in water while hydrophobic ("water-fearing") substances do not dissolve in water.

7. Water dissociation - water has a slight tendency to fall apart (dissociate) into hydrogen ions (H^+) and

hydroxide ions (OH^-). Other substances can also dissociate in water. An acid is a substance that gives off hydrogen ions when dissolved in water while a base is a substances that accepts hydrogen ions (thereby increasing the hydroxide ions). The concentration of hydrogen ions is very important to living cells; many processes will only occur at the proper rate under very specific hydrogen ion concentrations. We measure this hydrogen ion concentration by the pH scale. Pure water has a certain hydrogen ion concentration its pH is 7 (neutral). The pH scale goes from 0 (the most acidic; highest hydrogen ion concentration) to 14 (the most basic; least hydrogen ion concentration). The pH of most cells is between 6.5 and 7.5; it is at these concentrations that life processes occur at optimal speed.

Carbon

Carbon is the element that life on earth is based upon; its bonding versatility is its key characteristic, allowing it to accept 4 other atoms.

Most organic compounds possess a similar "backbone" of carbon atoms bonded together in chains or rings. The specific properties of the compounds are associated with functional groups that hang on the backbone.

Classes of Organic Molecules

1. *Carbohydrates* - the most abundant organic compounds. They function as energy sources for cells and as structural components of cells. The simplest carbohydrate is glucose - $C_6H_{12}O_6$.

   Monosaccharides are monomers or simple sugars. Examples are glucose, fructose, and galactose. All three have the chemical formula $C_6H_{12}O_6$, but they differ in their arrangement of the individual atoms and have different properties.

Disaccharides - are two simple sugars bonded together. Many plants typically store carbohydrates as disaccharides. Other examples are lactose or milk sugar and maltose, the sugar in barley used to make beer.

Polysaccharides - may consist of thousands of monomers of glucose or other simple sugars. Examples include starch (carbohydrate storage in plants), cellulose (a structural polysaccharide in plants and fungi), and fructans (storage products in leaves and stems). All of these are composed of repeated monomers of glucose.

2. *Lipids* - include fats, oils, waxes, steroids, and phospholipids. They function as an energy storage, as waterproof coatings, and as chemical messengers.

A. *Triglycerides* - fats and oils

(a) Saturated fats are those in which all the carbon atoms of the fatty acids are bonded to at least two hydrogen atoms. They form straight chain polymers in which the fatty acids are packed very closely and include fats such as bacon fat and lard.

(b) Unsaturated fats are those two adjacent carbon atoms on the fatty acids share a double bond and therefore have fewer hydrogens. They cannot pack as tightly as saturated fats. At normal temperatures they are liquids (These are the oils). Polyunsaturated fats have even more of these double bonds.

B. *Waxes* - insoluble in water, useful as waterproof coatings for organisms, as a structural component of cell walls.

C. *Phospholipids* - consist of only two fatty acids attached to a glycerol molecule and a phosphate group in place of the third fatty acid. The phosphate group is water soluble while the rest of the molecule is insoluble in water. The cell membranes surrounding every cell is made of a bilayer of phospholipids.

3. *Proteins* - The specialized shapes and functions of different cell types (something we will explore later in the course) depend upon the bewildering variety of proteins. Proteins have many roles in cells (and between cells). They include, but are not limited to:
    1. *Structural proteins* - form cell parts.
    2. *Regulatory proteins* - control cell processes.
    3. *Enzymes* - facilitate (help) many chemical reactions; they do this by lowering the amount of energy needed to start the reaction; the enzyme is not permanently altered in the process.
4. *Hormones* - chemical messengers.
5. *Transport proteins* - carry other substances around cells or from cell to cell.

Proteins consist of long chains of amino acids, the building blocks of proteinsroteins are made of one or more polypeptides. There are usually 100 to 10,000 amino acids in a typical protein molecule. Many millions of combinations of amino acids are possible.

Protein Structure - proteins have four levels of organization.

1. *Primary structure* - simply the order of amino acids in a polypeptide strand.
2. *Secondary structure* - regions of localized bending resulting in (a) an alpha helix and/or (b) pleating (like an accordion) called beta sheets give proteins their secondary structure. Weak hydrogen bonds cause this structure.
3. *Tertiary structure* - the three-dimensional folding of the entire polypeptide chain.
4. *Quaternary structure* - the fitting together of two or more polypeptide chains, thus forming a functional protein.

Nucleic Acids

Large organic molecules whose chief function is to carry the genetic information in the form of a code. There are two main types: Deoxyribonucleic acid (DNA) and ribonucleic acid (RNA). Nucleic acids are made up of polymers of nucleotides, their building blocks. DNA exists as a double helix while RNA is a single helix. See Structure of a single nucleotide in text.

In DNA the four different nitrogen bases are adenine, thymine, guanine, and cytosine. RNA substitute uracil for thymine, otherwise it is the same.

Another, very important molecule that is composed of nucleotides is adenosine triphosphate (ATP); it is technically not a nucleic acid such as DNA and RNA. It is known as the energy molecule; it provides immediate energy for the activities of every living cell. We will discuss this vital molecule in further detail when we discuss Cell Respiration.

CHAPTER 2

Measured Vs. Exact Numbers

Exact numbers are numbers that are exact by definition, such as:

1 inch = 2.54 cm exactly or

1 gallon = 231 cubic inches exactly or

1 foot = 12 inches exactly

or

numbers that come in integers and are not likely to be available in amounts smaller than integers.

When you ask for seating in a restaurant, the number of people is an integer, an exact number.

Measured numbers are an estimated amount, measured to a certain number of significant figures without the benefit of any natural unit or a number that comes from a mathematical operation such as averaging. You would get a measured number from using a meterstick to find the length of a board or using a graduated cylinder to find the volume of a liquid. Average numbers of people would make a measured number, even though people naturally come only in integers. The average family in the U.S. has 2.37 children.

Scientific Notation

There are many very large and very small numbers in scientific studies. How would you like have to calculate with:

1 mol = 602,200,000,000,000,000,000,000 atoms

or

1 Dalton = 0.000,000,000,000,000,000,000,00165 g?

You can streamline large or small numbers with scientific notation. The standard is that you place the decimal point after the first significant digit and adjust the exponent of ten so that there is no change in the value of the number. Think of the change as creating a new number with two parts, a digit part and an exponent part, from the old number. To put the decimal point behind the first digit, you must divide or multiply the original number by some integer power of ten. Then you must do the opposite (inverse) to the exponent part of the new expression so that there is no change in the value of the number.

(0.000,000,000,000,000,000,000,00165 x 10^{24}) 1/10^{24} = 1.65 x 10^{-24} or 1.65 E-24

(602,200,000,000,000,000,000,000/10^{23}) x 10^{23} = 6.022 10^{23} or 6.022 E23

The original numbers have the same value as the exponential forms, but the exponential forms have the decimal point in the right place. The 'right place' is to the right of the first digit. The 'E' in the number stands for exponent. Your scientific calculator will use the numbers in the shortened form, usually best represented by the 'E' form. Don't get caught making too much of this. You have seen it before. The number 'five point two million' is the same as 5.2 E6. The number of the power of ten only indicates how many places you need to move the decimal to get the long form of the number back. The only question you might have trouble with is which way to move the decimal. The easy way to remember that is: numbers that are less than one have negative exponent numbers in the scientific notation form, and numbers that are larger than one have positive exponent numbers. Very often chemistry professors will tell you they want answers in scientific notation if the number is larger than one thousand or smaller than one thousandth. Keep your professor happy. Find out exactly what is required in your course and follow the instructions to the letter. Get some practice with scientific notation in the Chemtutor Quickquiz.

The table shows the fully written out number, the way we say the number in English, and the way to write the number in scientific notation. The numbers in scientific notation are to two significant digits.

Full number	Words for number	Sci. notation styles
5,000,000,000	five billion (USA)	5.0×10^9 or 5.0 E9
500,000,000	five hundred million	5.0×10^8 or 5.0 E8
50,000,000	fifty million	5.0×10^7 or 5.0 E7
5,000,000	five million	5.0×10^6 or 5.0 E6
500,000	five hundred thousand	5.0×10^5 or 5.0 E5
50,000	fifty thousand	5.0×10^4 or 5.0 E4
5,000	five thousand	5.0×10^3 or 5.0 E3
500	five hundred	5.0×10^2 or 5.0 E2
50	fifty	5.0×10^1 or 5.0 E1
5	five	5.0×10^0 or 5.0 E0
0.5	five tenths	5.0×10^{-1} or 5.0 E-1
0.05	five hundredths	5.0×10^{-2} or 5.0 E-2
0.005	five thousandths	5.0×10^{-3} or 5.0 E-3
0.000,5	five ten-thousandths	5.0×10^{-4} or 5.0 E-4
0.000,05	five hundred-thousandths	5.0×10^{-5} or 5.0 E-5
0.000,005	five millionths	5.0×10^{-6} or 5.0 E-6
0.000,000,5	five ten-millionths	5.0×10^{-7} or 5.0 E-7
0.000,000,05	five hundred-millionths	5.0×10^{-8} or 5.0 E-8
0.000,000,005	five billionths	5.0×10^{-9} or 5.0 E-9

Significance and Rounding

All the numbers in the above table have only two significant digits. Only the five and zero in each number has any numerical meaning other than place-holding. Let's say that Delhi, India has five million people. We have expressed that number in one significant digit. That might be sufficient for such a number. Are we considering just people within the city limits? How about people who live outside the city and come in only for business? What year are we specifying? Let's say we have enough information to number the population of Delhi at 5.1 million. That is now more a more accurate number that claims a two digit significance. What if we were to go absolutely wild and say that Delhi has 5.1376504×10^6 people? That is the same number, isn't it? First, it ridiculously claims eight significant digits. Even worse, that figures to 5,137,650.4 people, and people just don't come in four-tenths of a person. Yes, you can claim to be far too accurate. With that in mind, here are the rules for considering the number of significant digits:

1. A and 9 claims significance.
2. All leading and following zeros that are only place-holders are not significant. The two numbers given as examples have a large number of merely magnitude-indicating zeros.
3. All zeros between two other digits are significant. The number 6.023 has a significant zero for a total of four significantdigits.
4. All zeros to the right of the decimal and to the right of other digits are significant. For instance, the number 43.500 has five significant digits, two of which are zeros.

You may hear the phrases significant digit, significant numeral, or significant figure to describe this idea. Chemtutor will sometimes shorten it to "sig figs."

We can round numbers to the proper number of significant digits by lopping off all digits past the number needed if less than five and rounding up the last needed digit if the following digit is five or more. For examples, here rounding to three significant digits:

3.4848 becomes 3.48;

4.1550 becomes 4.16;

5,786,899 becomes 5,790,000; and

0.000,347,00 becomes 0.000,347.

How do you know when you need to round? In multiplication and division, the answer cannot have more significant digits than the number with the smallest number of significant digits used to calculate the answer. So a four significant digit number multiplied or divided by an eight significant digit number will result in a number that can only claim four significant digits. Wisconsin has 379 cities with about 5.1 thousand people in it. How many people live in all these small towns? 379 × 5.1 E3 = 1.9329 E6, but the answer can only claim two significant digits. The answer must be 1.9 E6 people because the number 5.1 E3 only has two significant digits.

How do you know where to round in actual measurements? The last digit is the one we get by estimation. For instance, if you have a graduated cylinder marked in milliliters and tenths of a millilitre, you should be able to estimate between the lines (interpolate) of tenths of millilitres and measure hundredths of millilitres.

The allowed significance works differently with adding and subtracting. The addition of a family of four to a city of three million does not significantly change the population of the city. If you have 1,578,000 chickens and you add 2,717 chickens to them, you have 1,581,000 chickens. Align these numbers one on top of the other so you can more easily see the reasoning behind this. No answer in subtraction or addition can have significant digits in Columns in which on the right there is not a significant digit in each participant number.

It is just as important to know when to round as how to round. In any math problem you should wait until the end to round; Only the final answer should be rounded. Carry as many significant digits as you can throughout the problem. On a calculator, the most efficient way to carry the maximum is to do all the calculation on the calculator. Arrange the problem so that you do not have to copy an intermediate answer only to re-enter it into the calculator. If you do find yourself needing to save numbers outside the calculator, copy several more significant digits than you think you need.

For help in significant digits and scientific notation, ask Chemtutor for the Quickquiz. For a good scientific calculator on the web, click here.

Back to the beginning of Numbers and Math Operations.

Dimensional Analysis

Dimensional analysis is a system of using the units of quantities to guide the mathematical operations. There are other names for the very same idea, for instance, unit conversion or factor label or factor-unit system. Here we will abbreviate it to DA. It can be used for conversions, to check work done with formulas, or, in uncomplicated problems, instead of formulas. As we will use it, DA can do much of the mathematical work of chemistry. We will also be doing some more complex math problems for which the use of formulas is necessary. The general rule for whether to use DA or W5P is: if the answer is the same dimension as the given units, you will likely use DA, but a formula best accounts for any change in dimension. (W5P method. See below.) To use the DA system for conversions, you should check to see that the unit given and the unit into which you want to change are the same dimension. To someone unfamiliar with that idea, the prospect of finding out how many feet are in 2.5 acres almost seems a reasonable one.

Begin with the known quantity. Place all the known quantity in the numerator of the beginning fraction if there is no denominator. To use DA, one must know the dimensions of units and conversion factors. Definitions, such as those found in the metric prefix and unit sheet can serve as conversion factors by dividing one side of an equation by the other, thus: since 1 mile = 5280 ft.

$$\left(\frac{1\text{ mile}}{1\text{ mile}}\right)=\left(\frac{5280\text{ ft.}}{1\text{ mile}}\right)\text{ so }\left(\frac{5280\text{ ft.}}{1\text{ mile}}\right)=1\text{ and }\left(\frac{1\text{ mile}}{5280\text{ ft.}}\right)=1$$

We can multiply any quantity by anything equal to 1 without changing the value of the quantity. Therefore, 5280 ft/mile or 1 mile/5280 ft is a conversion factor for length. These conversion factors can change 6.20 miles to a number of metres in serial fashion thus:

$$\left(\frac{6.20}{1}\right)\left(\frac{5280}{1}\right)(12)(2.54)\left(\frac{1}{100}\right)=9978\text{ metres}$$

Notice you can visualize the definitions you know from the units-and-definitions section in each one of the conversion factors. The definition 1 mile = 5280 ft can convert miles to feet. We want to cancel out the miles, so we place 1 mile under the 5280 ft. Now the mile units can cancel with one in the denominator and the other in the numerator. If we were to stop at this point, the answer would be in units of feet. Similarly, use 1 foot = 12 in. to go from feet to inches, placing the foot unit in the denominator to cancel with the foot unit in the numerator. The definitions and conversions in the units and measures and definitions chapter become very convenient for use as conversion factors. It is necessary to know them by rote to be able to easily use the system. If you know the definitions well, not only will you escape having to look up the right definition, but you will much more easily spot the best way to convert the numbers. The metric system definitions come from the powers of ten of the metric prefixes. The metric 'staircase' method allows for changes from any magnitude to another in one conversion step by 'counting the steps up or down the staircase.'

We can use the density of a material as a conversion factor between mass and volume rather than a formula in this manner: The density of mercury is 13.6 g/cc. What is the mass of two liters of mercury?

GIVEN	DACF	DENSITY	RAW ANSWER	DACF	ANSWER
$\left(\frac{2.00}{1}\right)$	(1000)	(13.6)	$= 27200$	$\left(\frac{kg}{1000}\right)$	$= 27.2$ kg

This math could have been done without stopping to calculate the intermediate answer of 27200 g, but the math always works, no matter in what order it is done. In both cases we carefully accounted for the units by canceling. A stepwise orderly progression of easily remembered changes can convert units to anything you need. The metric system is particularly easy to use with DA. The definitions are multiples of ten and need scrupulous care to stay untangled. Here is a way to think of it using a change from kilometers to millimetres. Kilometres being the larger unit, begin with one of the larger unit. The number of smaller units (mm), is the power of ten that is the number of steps up the metric staircase. This process keeps the exponent positive.

1 km = 1×10^6 mm or 1 km = E6 mm

Problems that involve many substances or a lot of addition and subtraction can become difficult with DA, but for conversions within a dimension or simple multiplication or division problems, even in serial fashion, it is a powerful tool. We can use DA when working formula problems by consistently using and canceling the appropriate units. The answer must come out in the proper dimension according to the units, or you should suspect something wrong. You may then easily convert the answer to units you want by using simple definitions as conversion factors. As you may have suspected, DA has many uses in the study of chemistry. Many common daily-living problems could be helped by thinking in DA.

Problem Solving Techniques - The W5P Method

One of the really difficult things to learn or teach in any science is the way to do word problems. For some students, there seems to be a mental block about it. Chemtutor teaches a method that may work for you, but the success of the method depends upon the background knowledge of the student in the units, dimensions, and formulas.

W5P Method (Wild and Wonderful Way to Work Word Problems)

Given - List all pertinent information with dimension symbol, number and unit.

Find - List the dimension of the quantity requested in problem.

Formula - With the dimensions in given and find, list the formula of formulas that fit.

Solve - Solve the formula for what you are looking for (find), substitute the number values in given, and perform the math on both the units and the numbers.

Answer - Check the answer for likeliness, make sure the units are appropriate, express the answer in scientific notation and to the accuracy required, and draw a box around it so it is obvious which number your answer is.

Here is a sample problem using W5P method. Lead is 11.3 g/cc. What is the volume of 24.5 kg of lead? If you recognize the units of density, you can immediately spot that D = 11.3 g/cc. The symbol of the dimension, density, is D. We show the number and units of the density with the symbol. The mass is 24.5 kg. (m = 24.5 kg.) Find the volume (V). The formula that has all the dimensions needed is the density formula from the formula list. We write it here in its original (memorized!) form. Solve for what you are looking for (volume here) using just a little bit of good old basic algebra. You would like to get the equation in the condition where you have V on one side all by itself. Do the algebra as stepwise as you need to in order to show yourself you have actually done it right. Next, substitute the measurements for the dimension symbols. Note that the kilogram and cc and gram units cancel, along with the number one thousand in the math. All we have left is to do the division and processing of the answer. The answer calls for three significant digits. The number does not need scientific notation because the number is less than one thousand and more than one thousand.

Given: D = 11.3 g/cc m = 42.5 kg Find: V

Original Formula	**Steps in Solving for "V"**	**Solved Formula**
$D = \frac{m}{V}$	$D(V) = m \quad V = \frac{m}{(D)}$	$V = \frac{m}{D}$

Substitute	**DA to Clean up**	**Raw Answer**	**Answer**
$V = \frac{24.5}{11.3}$	$1000 \; \frac{\text{litre}}{1000}$	= 2.168 L =	2.17 L

Notice that the "given" section includes the symbol for the dimension of the information, the number and unit. "Find" shows the dimension of the information requested. The formula is not labeled as such, but it is the relationship that has the requested information among the dimensions and all other dimensions in the formula are known. (Here we are looking for V and we know D and m.) The next step is to solve the formula for the dimension we need. Then substitute the given quantities for the known dimensions and process the answer.

Here is an opportunity to show you a small but useful way to save yourself some grief. Consider the unit g/cc in the above math. The unit is a fraction in the denominator. It is easier to think of as cc/g in the numerator. (The rules of math permit you to move a fraction from the denominator to the numerator if you invert the fraction.)

$$\overset{\textbf{A}}{\frac{24.5\text{ kg.}}{11.3\left(\frac{\text{g}}{\text{cc}}\right)}} = \overset{\textbf{B}}{\frac{24.5\text{ kg.}\left(\frac{\text{cc}}{\text{g}}\right)}{11.3}} = \overset{\textbf{C}}{\frac{24.5\text{ (cc)}}{11.3}} = \overset{\textbf{D}}{\frac{24.5\text{ (k)cc}}{11.3}} = \overset{\textbf{E}}{\frac{24.5\text{ litres}}{11.3}}$$

Fraction A comes from the original substitution of the numbers and units into the solved equation. Fraction B shows the inversion of the fractional unit into the numerator. This is a really useful way to simplify fractions that seem complex. Fraction C shows the same fractional unit integrated into the whole fraction and cancels the unit of gram. Step C is not needed if you want to cancel the gram unit in the numerator. Fraction D shows what happens when you cancel a unit without its metric prefix. The k is just one thousand now, so k (cc) is the same as one thousand times cc which is a liter. Fraction E shows the final unit. Many students make math mistakes by failing to simplify fractions in the denominator.

Back to the beginning of Numbers and Math Operations.

The Temperature Box

There are two types of common conversion that are not easily fit for calculation by DA because they require formulas among the different units. Temperature conversions are among the units Celsius, Kelvin, Fahrenheit, and Rankine. Acidity units are pH, hydrogen ion concentration, hydroxide ion concentration, and pOH. A conversion box is a fine teaching device for this type of unit association group.

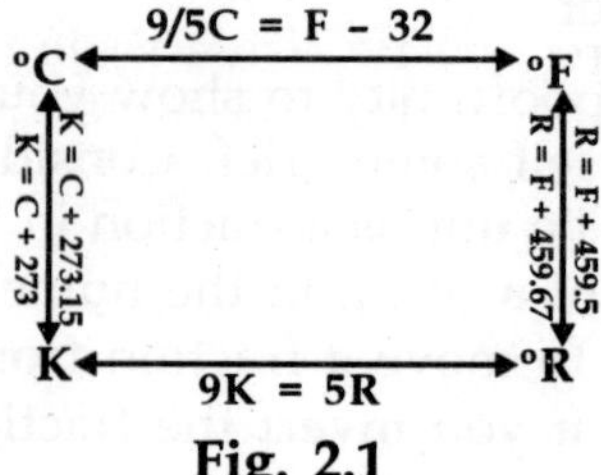

Fig. 2.1

(For going around the box, use K = C + 273.15 and R = F + 459.67 to get more accurate results.)

Some common temperatures in all four scales to the nearest whole degree.

There are only three temperature measurement scales commonly used. The Kelvin, Fahrenheit, and Celsius or Centigrade scales all have common uses. The Rankine scale is not commonly used, but it is needed here for the learning experience. Consider four identical liquid-type thermometers side-by-side with different markings on them. This is a good mental image because These four scales measure exactly the same thing. The small superscript 'zero' before the symbol means 'degree.' Some people make a big issue of omitting the degree sign before Kelvin. They say 'Kelvins' instead of 'degrees Kelvin.'

The Kelvin and Rankine scales are both absolute temperature scales. Since temperature is a measure of the average velocity of the atoms or molecules in a material, it makes sense that there is a temperature that represents the condition of no molecular motion. That temperature is called 'absolute zero'. Both Kelvin and Rankine scales are absolute scales. The zero mark on these scales is at absolute zero. There are no negative temperatures on either scale. Since they have the same zero point, the relationship between them is a simple proportion. You will remember the formula as the Dog Equation - - Canine equals Arf-ive. Yes, K 9 = R 5 or, as you more commonly see it, 9 K = 5 R. No applause, please.

The size of the Celsius degree is the same as the size of the Kelvin degree. The Celsius scale just has its zero at the freezing point (or melting point, for they are the same) of water, 273.15 K. The size of the Fahrenheit degree is the same as the size of the Rankine degree. To calculate the temperature in Rankine, add 459.67 to the Fahrenheit temperature. For instance, the freezing point of water is 32 °F or 491.67 °R.

The easy way to be sure to know the conversion between Fahrenheit and Celsius is as follows:

(a) Only remember one formula;

(b) From the Dog Equation the proportion of Celsius and Fahrenheit is nine to five. 9/5 C must be the right proportion; and

(c) Fahrenheit zero is 32 degrees below the freezing point of water.

Therefore, the equation is:

$$9/5\ C = F - 32$$

Know This

This is the only equation you will need or want. Some texts will ask you to memorize two equations, one solved for Fahrenheit and the other solved for Celsius. To get Celsius temperature from Fahrenheit, you must solve for 'C,' but, for the convenience of remembering only one easily related equation, the price is cheap.

The temperature box is a device that can give you as much practice as you need. Here is how to use it. Choose a temperature, for instance, 37 °C, normal human body temperature. Calculate around the box either clockwise or counter-clockwise. You should return to the same Celsius temperature. For rough one-time calculations, you can use the more generally quoted change between Celsius and Fahrenheit and between Fahrenheit and Rankine. For going around the temperature box, you should use the more exact numbers in the parentheses.

Percentage Calculations

Converting raw numbers to percentages is easy once the parts are defined. A percentage is the target over the total multiplied by one hundred per cent.

$$\frac{\text{Target}}{\text{Total}} \times 100\% = \text{Per cent}$$

There are thirty people in the classroom. Of them, seventeen are male. What percentage of males are in the classroom? 'Seventeen males' is the target we have defined. 'Thirty people' is the total. Seventeen divided by thirty times one hundred is 56.66667 to seven significant digits or 56.7 to three significant digits. Males are people, so we cancel the units. The answer is 56.7 per cent.

$$\frac{17 \text{ males}}{30 \text{ people}} \times 100\% = 56.66667\% = 56.7\%$$

In many cases, the most difficult part of using percentages is identifying the target and the total. Percentages do not have any other unit attached to them other than the per cent. After dividing one unit by the same type of unit and cancelling the units, that makes sense.

Basic Algebra

Algebra operations can be performed on equations with symbols, numbers, or measurements. An equation states that the right side of the equation is equal to the left side. The rules of manipulation of the two sides of an equation without changing the equation are:

(a) you can multiply or divide both sides by the same thing;

(b) you can add or subtract the same thing from both sides;

(c) you can raise both sides to the same power, you can change both sides to or from any exponent.

Also, you can substitute anything in an equation for something of equal value.

One of the really big mistakes many novice calculators make is attempting to do too much between the ears before enough practice has been done. The best advice is that calculating with a pencil (or writing implement of any type) is much more reliable than calculating mentally. You may have heard if 'back- of- the- envelope' calculations by scientists or engineers. Most people of science or technology understand that even something scratched on the back of an envelope is more reliable than mental calculation. Show your work, if not to your teacher, to yourself. Show the addition of the same thing to both sides. Show the cancellation of units. Show the stepwise conversion of what you know to what you need. Except for log and antilog work for pH calculations, the use of more complex algibraic operations in basic chemistry course is rare. The use of quadratics or complex factoring is rare.

Proportionality - Direct Proportion

A boy and a girl are in the back seat of the car on a long trip through the countryside. Their mother suggests that they count the cows on their side of the car to keep them occupied. The boy sees a large number of cows on his side and begins furiously counting, "1, 2, 3, 4, 5, . . ." It is obvious from his disappointment that he did not get a chance to count all his cows. Then a much larger herd appears on the girl's side. She looks at the herd and proudly announces, "Six hundred and seventy seven cows." The boy angrily says, "Oh, you couldn't have counted that many so quickly." She retorts, "It's easy. All you have to do is count the legs and divide by four."

Mathematically, of course, she is correct. The number of cows is proportional to the number of legs; as the number of cows increases, the number of legs increases and vice-versa. This could be expressed mathematically as:

number of cows   number of cows legs

You would read the proportionality as:

"The number of cows is proportional to the number of cow legs."

An example in chemistry of proportionality is the relationship of the pressure and temperature of the same gas at a constant volume. When the temperature increases, the pressure increases. When the temperature decreases, the pressure decreases. The formula for that can be expressed:

$$\frac{T_1}{T_2} = \frac{P_1}{P_2}$$

Proportionality - Inverse Proportion

Consider the table of hydrogen ion concentrations versus hydroxide concentrations. As the $[H^+]$ increases, the $[OH^-]$ decreases. As the $[H^+]$ decreases, the $[OH^-]$ increases. This relationship is inverse proportionality.

It is something of a joke in corporate circles that the importance of a person is inversely proportional to the number of keys carried. The CEO of the company (of 'infinite importance') does not have to keep any keys because all doors are opened for the CEO. The janitor (lowest in importance) has a large ring of keys. This can be shown in the proportion manner by:

$$\text{importance} \propto \frac{1}{\text{number of keys}}$$

Using the pressure and volume of the same gas at constant temperature, inverse proportionality can be expressed mathematically by:

$$P_1V_1 = P_2V_2$$

How Formulas show Direct and Inverse Proportion

It is easy to spot proportionalities in mathematical formulas. Let's use the universal gas law, PV = nRT, as an example. If the two variables are both in the numerator (or both in the denominator) on opposite sides of the equation, as P and T in the universal gas law, they are proportional. It two variables are on the same side of an equation, they are inversely proportional, as in P and V in the same equation.

For an explanation and discussion of each problem, click on the question mark following the question. A set of answers follow each problem set.

These problems may not have much to do directly with chemistry, but they are designed to help you understand the methods of thinking about and solving problems in chemistry.

Problem Set # 1 – Conversions Using DA

Do these Conversions in the Dimensional Analysis Style. Answer in Scientific Notation if > E3 or < E-3.

1. What is 1.50 mm in km?
2. How many nanoseconds are in 1.50 days?
3. How many floz. of water do you have in exactly half of a ten gallon can?
4. A car is going 60.0 MPH. How fast is that in ft/sec?
5. A car is going 62.0 MPH. How fast is that in KPH?
6. How many mm^2 are there in one square kilometer?
7. How many in^2 are there in an acre?
8. How many mm^2 are there in 0.550 acre?
9. A car goes 0 to 60.0 MPH in 5.00 sec. Write that acceleration in m/sec^2? [Start with (60 mi)/(hr.)(5 sec)]
10. Light travels at 3.00 E8 m/sec. How fast is that in MPH?
11. An elephront has a mass of 1.80 tonnes. What is its weight in ounces?
12. Mercury has a density of 13.6 g/cc. What is that in #/gal?
13. A light year is the distance that light goes in a year. Using data from # 10, how long is a light year in miles ? (Rate times time = distance).

CHAPTER 3

Units, Measures and Dimensions

Measurements

Measurement is the most useful form of description in science. Often the most useful measurements are those that have a number and a unit, such as '12.7 inches'. Here '12.7' is the number and 'inches' is the unit. This unit of inches in the example is one of the common units in the dimension of length. A number, then, is an expression in numerals. A unit is a recognized way to divide the essence of a dimension for measurement, and a dimension is a measurable physical idea. Here is a bit of advice you can overlook only at your peril: To become fluent in the subject you should memorize the basic background of information. The following units, dimensions, and measures are so basic to the study of Chemistry that you could always help yourself by memorizing these. Chemtutor offers a Quickquiz on this table along with the metric prefixes. The real test of whether you know this well enough is to recognize the dimensions of any measurement and know its symbol and magnitude from the unit alone. Chemtutor Quickquiz will help you with that.

Table 3.1: Dimensions and symbols

Dimension	Symbol	Metric Units	Symbol	"English" Units
Length	S, l, d, r	meter (+m.p)	m	Ft, in, Yd, mi, etc.
Area	A	Sq. meter, etc., hectare	m^2	sq. Ft. etc., acre
Volume	V	cu.meter, etc., liter	m^3, L	cu. Ft. cu. in. etc., gal, Floz
Time	t	sec (+m.p) sec., min., hr., day, yr., etc. (both metric & English)		
Mass	m	Kilogram (+ m.p.), AMU	kg	(slug, rarely used)
Force (weight)	F, F_w	Newton (+m.p.)	N	Pound (#), Oz, etc.
Velocity	v	meter/sec, KPH, etc	m/sec	Ft/sec, MPH, etc.
Acceleration	a	meter/sec.sq., etc.	m/sec^2	Ft/sec sq., etc.
Pressure	p	N/sq.m, atm., Pa	atm, Pa*	#sq.in (PHI), inHg. etc.
Density	D	g/cc, Kg/liter, etc.	g/cc	#/cu.Ft,#/gal, etc.
Temperature	T	Celsius or Kelvin	°C	Fahrenheit or Rankine

(Contd...)

Dimension	Symbol	Metric Units	Symbol	"English" Units
Energy	E	Joule (+ m.p.)	J	foot-pound
Heat	Q	calorie (+ m.p.)	cal	BTU (British Thermal Unit)
Concentration	C**	gram/L, mol/L, Molar	M	(#/gal or #/cu.ft, rare)

Abbreviations: Ft = foot; in = inch; AMU = atomic mass unit; KPH = kilometers per hour; MPH = miles per hour; gal = gallon; PSI = pounds per square inch; cc = cubic centimeter; inHg = inches of mercury; Pa = Pascal; m.p. = metric prefixes; cu. = cubic; sq. = square; atm = atmosphere.

* The unit Pa, for Pascal, is a unit of pressure that is the standard unit for the SI system, the MKS system in the metric measurements. The unit of Pascal, however, is rarely used in chemistry. Instead, the unit "atm," for "atmosphere," is still most used in chemistry.

** The symbol "B" is now the official symbol for concentration in the SI, but there are still chemistry texts using the "C" as is shown here."

Dimensions, Units and Symbols

Notice the symbols of the dimensions as they would be used in formulas. The basic metric symbol or the symbol of the most used metric unit is listed after the metric units.

The Table 3.1 lists almost all the dimensions you will need in this course, the symbol for each dimension as it will be used in common formulas, and the units of each dimension. Notice Chemtutor has two systems of measurement displayed that you should know. There are really two commonly used metric subsystems. Most chemistry texts will use the MKS system (meter, kilogram, second) rather than the less-used CGS (centimeter, gram, second) system. A *system* is defined by its basic measure of distance, mass, and time.We will use the MKS system, also called the S.I., or International System. The symbol for only the basic unit of each dimension in the metric system is on the list.

Metric System Vs. "English System"

The metric system typically uses only one root word for any basic dimension such as for length, the meter. All the metric units of length use the root word 'meter' with the metric prefixes in the next table. Our common system in the United States is not really a system, but is a thrown-together mess of measurements with no overriding order. Chemtutor, as does most of the United States, calls this group of measurements the "English system". While calling it that is a considerable slander on the English people, the United States and Liberia are the only nations on earth to still cling to it. Chemtutor thinks that the English system makes a fine learning tool, along with being wonderfully poetic. You will want to know how to relate the English System to the metric system. Particularly notice the large number of units of length in the English system. This is only a small number of the common ones. We regularly use fathoms to measure depth in water and furlongs to measure distance in horse racing. There are many little-used English length units such as the barleycorn (one third of an inch) that may be picturesque, but are not used today. Notice that we define the barleycorn as a third of an inch. The way to relate one English unit to another is by definition. Length is the most common measurement. As a result, it has not only the largest number of words to describe it, but it also has the largest number of symbols to represent it in formulas. The English language also uses distance, long, width, height, radius, displacement, offset, and other words for length, sometimes in specialized applications.

Length

A meter is a little longer than a yard, so a meterstick that has inches on the back of it will has just a bit over 39 inches on the English side. Typically, on the English side, the inches are broken into halves, fourths, eighths, and perhaps sixteenths. On the metric side one meter breaks down into ten decimeters, one hundred centimeters, and a thousand millimeters.

Area

An area is a length multiplied by a length. (A= l l as in the formula list.) An area is an amount of surface. Almost all area units are length units squared, such as: square meter (m^2), square centimeter (cm^2), square inch (in^2), etc. The acre and hectare, units of land measurement, are the only units commonly used that are not in the 'distance squared' area unit format. An acre is defined as 43,560 square feet, so in using the unit 'acre' in dimensional analysis, the definition can be used to relate the acre to other units. Notice the squaring of a unit of length. A meter multiplied by a meter is a square meter. A foot by a foot is a square foot, etc.

Volume

Volume is length multiplied by length multiplied by length. You may have heard that volume is length times height times width, but it means the same thing. (V= l l l)

You may think of a volume as the space inside a rectangular (block-shaped) fish tank. Volume is the measure of an amount of space in three dimensions. Because volume is such a common type of measurement, it is unique in that it has two types of commonly used root word in both metric and English systems. The metric roots are liter and cubic meter. The English system also uses cubic length and an extensive array of units that are not in the cubed length format. Again, analagously to area measurements, a cubic meter is a meter multiplied by a meter multiplied by a meter, and a cubic foot is a foot by a foot by a foot.

Time

Time is also a bit odd in its units. In both systems the units of less than a second are in the metric style with prefixes before the second. Time units of more than a year are in a type of metric configuration because they are in multiples of ten (Decades, centuries, millennia, etc.). The dimension of time is messy for good reason. The more commonly used time units from day to year are all dependent upon the movement of the earth. The unit of 'month,' particularly if it is directly related to the moon, is useless as an accurate unit because it does not come out even in anything. Having 60 seconds in an hour and twenty-four hours in a day come about from the ease of producing mechanical clocks (Is it time to switch to metric time? How would you like, say, ten hours in a day, one hundred minutes in an hour, and one hundred seconds in a minute. It would come out to almost the same length of second.).

Mass

Mass is an amount of matter. Mass has inertia, which is the tendency of matter to stay where it is if it is not moving, or to keep moving at the same rate and direction if it is already moving. You could measure mass by an inertial massometer. Visualize a metal strip held tightly on one end and "twanged", or given a push to make it vibrate on the other end. It has a natural pitch to vibrate. If you were to put a mass on the end of that strip, you would change the pitch of the vibration. The change of pitch would make it possible to calculate the mass of the added object. This measurement of mass is completely independent of gravity, the way we often weigh a mass by comparing the push or force of the mass on a surface. Mass is a more accurate way of thinking of amount of matter compared to weight. The metric system is mass-based whereas the English system thinks in weight. Consider that an astronaut in near earth orbit has no weight because the gravitational attraction cancels inertia, but the mass of the astronaut remains the same. The metric root word of mass is the gram. Notice the difference between the 'root word,' gram, which is the basis for adding metric prefixes, and the system base of kilogram, the mass unit of the S.I. metric system.

Force

A force is a push or a pull. Those simple words are the best definition of a force under our limited experience. A force can not be seen or heard directly, so it is a bit of a difficult concept beyond the simple definition. Having basic metric units like 'kilogram-metre per second squared' make the idea of force hard to think about using that tool also. It is shameful to give you this in the same manner as a British sex education, but until a better way comes about, that's all there is to be said about it.

Weight

Weight is a downward force due to the mass of an object and the acceleration of gravity. The English system can conveniently use the idea of weight to measure amount of material because there is very little difference in the acceleration of gravity over the surface of the earth. There are certainly other forces besides gravity. Magnetism produces a force. Electric charge produces a force.

Velocity

Velocity is a complex dimension. The unit of velocity is a combination of more than one type of basic dimension. A velocity is a distance per time. The word 'per' here means 'divided by,' and distance divided by time is not only the definition of velocity, but it is the easy way to remember the velocity formula, $v = d/t$. Velocity also has the name of rate. You might know the same formula as, 'rate times time equals distance.' Here's where we could start complicating the math by using calculus, but we won't. If you are taking a course that requires calculus, the math is only slightly different, but the basic ideas behind it are the same.

Acceleration

An acceleration is just another step down the same road as velocity, that is, acceleration is a distance per time per time, or, another way to see it, distance per time squared. An acceleration is a time rate of change of velocity. If something changes its velocity, it has an acceleration. An acceleration causes an increase or decrease in speed or a change in direction. Newton and Einstein identified gravity as an acceleration. Gravity has a fairly consistent amount of acceleration on the surface of the earth, that is 32 ft./sec^2 or 9.8 m/sec^2. As you can see, the acceleration of gravity, 'g,' can substitute for the 'a' of acceleration in the formulas below when the acceleration is due to gravity.

Pressure

A pressure is a force per area. You can almost see the pressure of the wind on a sail. The pressure of the wind is the same, so the larger the area of the sail, the greater the force of the wind on the ship. Pressure unit definitions that we need for this course revolve around the unit 'atmosphere' because historically the pressure was first measured for weather.

Density

Density is mass per volume, weight per volume, or specific gravity, which is the density of a material per the density of water. Metric system densities are usually in the units of mass per volume, such as kg/L (kilogram per litre) or g/cm^3 (gram per cubic centimeter). English densities are usually in weight per volume, such as #/gal. (pounds per gallon) or #/ft^3 (pound per cubic foot). Specific gravity has no units (!) because it is a comparative measurement. Specific gravity is the density of a material compared to the density of water. Expressing density as specific gravity shows neither system.

We can have fun in a density demonstration by passing a large-grapefruit-sized ball of lead around the class. That size of lead ball weighs about 35 pounds. People do not expect something that compact to weigh so much. One way to think of density is, 'How much mass is packed into a volume.'

Temperature

Temperature is a bit more subtle dimension. What we really measure is the average velocity of the atoms or molecules in the material. One way to measure it is by the expansion of a liquid in a very small tube. This is the shape of a liquid (usually mercury or alcohol) in a thermometer. The Fahrenheit scale is still not a bad one for use with weather. Scientists are more likely to use the Celsius or Centigrade scale. Gas law calculations require the Kelvin scale because it is an absolute scale. The other absolute scale, Rankine (pronounced "rank-in"), is useful for teaching purposes, but is not in common use.

Energy

Energy is the ability to do work. A Joule, the metric unit of energy is a kilogram-metre-square-per-second-square. Both of those ideas can be difficult to wrap your mind around. The easier way to think of energy is perhaps by its various types. You should have an intuitive feeling that a 50 pound rock held above your head has more energy of position in a gravitational field than the same 50 pound rock by your feet. A rubber band pulled back has more spring energy than a lax one. A speeding train has more energy of movement than a still one. We usually value petroleum not for its beauty, but for its chemical energy content. Energy is transferable from one type to another, but is not lost or gained in changes.

Heat

Heat is a form of energy. It is the energy of the motion of molecules. Even though heat and energy are fundamentally the same dimension, we measure and calculate them differently. We define a calorie (note the lower-case 'c') as the amount of heat that increases the temperature of a gram of liquid water one degree C. The BTU, the English unit of heat, is the amount of heat that increases the temperature of a pound of liquid water one degree F. A food Calorie (note upper case 'C') is one thousand heat calories of usable food energy. That is, the food Calorie reflects the type of living thing eating and using the energy. So the food Calorie depends on the type of (animal) eating it. A cow or a termite could get much more food value from a head of lettuce than a human being can, so what is a Calorie for us would be different for them.

Concentration

Concentration is amount of material in a volume. In this course, we will stay mostly with measuring the amount of solute in a solution. There is more on this in the chapter on solutions, and we really need to explain the idea of mol or mole before a thorough explanation of concentration can mean much.

Notice the formulas in the table below. Some of the simple ones we use in this course only for practice with problem-solving techniques and for defining the units and dimensions. There are a few items in the formulas that have not been mentioned yet, such as c, the specific heat; n, the number of mols; and R, the universal gas constant. These we will consider in context as we use them.

Formulas

A = l l V = l l l V = A l v t = d F = m a (Fw = m g)
a = v/t a = d/t2 P = F/A C V = n D = m/V (D =Fw/V)
P V = n R T Q = m c∇T
Circle Area, Ac = πr^2.
Cylinder Volume = Vc = Ac l = πr^2 l

A formula is a relationship among dimensions. The symbols for the dimensions in the formula list are in the dimension list. Note the capitalization or lack of it in the symbols, for instance, V = volume and v = velocity; C = concentration and c = specific heat, etc. Also, there are some letters written after and slightly under a symbol called a subscript. Subscripts indicate a special case of the symbol, as you see above with the area of a circle being represented by the A for area and a subscript c for circle.

Definitions to Change Units

There are three types of definitions you should know for changing units, English system definitions, metric system definitions, and changeover definitions between the two systems.

There are a small number of English system definitions listed that you should know by rote. Notice that we take the same approach here with one of the larger unit being stated first and then some number greater than one of the smaller unit. All of these English definitions are exact definitions except for the cubic feet-to-gallons relationship.

Take a look at any edition of the Chemical Rubber Company (CRC) Handbook of Physics and Chemistry and you will see the incredible number of non-metric units.

English System Definitions you should know by Rote

1 ft.	=	12 in.	1 mi.	=	5280 ft.	1 cup	=	8 Floz.
1 pint	=	2 cups	1 qt.	=	2 pints	1 gal.	=	4 qts.
1 #	=	16 Oz.	1 ton	=	2000 #	1 acre	=	43560 ft^2
*1 ft^3	=	7.48 gal.						
1 gal.	=	231 in^3						

Metric Prefixes as Factors of Ten

FACTOR	PREFIX	SYMBOL
+18	exa	E
+15	peta	P
+12	tera	T
+9	giga	G
+6	mega	M
+3	kilo	k
+2	hecto	h
+1	deka	da
0	*Root word only*	
-1	deci	d
-2	centi	c
-3	milli	m
-6	micro	μ
-9	nano	n
-12	pico	p
-15	femto	f
-18	atto	a

The table (*page 48*) includes only the commonly used metric prefixes. There have been some metric prefixes suggested for some of the exponents of ten not listed here, but they are not in common use, or are in use by only a small number of people for limited use. The prefix "myria-" (my or ma) as E4 is a good example. The word "myriad" means ten-thousand, so the prefix is well documented in language.

A Few Odd Metric Definitions

1 metric tonne	= E3 kg	1 mL.= 1 cc	= 1 cm^3
1 Ångstrom	= E-10 m	1 cubic meter	= 1000 L

Metric System Definitions

Metric system definitions are relationships between units with the same rootword that only the prefix changes. The Metric Stair-nano -9 case is just a way to visualize the relationships on the metric prefixes. We make a metric system definition in the following way, using the units kilometer and millimeter as an example:

1. Pick the largest metric prefix. Begin the metric definition with of the larger units, e.g. 1 km = (some number of) millimeters.
2. Count the number of 'steps' down the metric staircase between the two metric prefixes. For instance, kilo- to milli- is six steps.
3. The number of the smaller unit is ten to the power of the number of steps between the metric prefixes. In our example.

   1 km = 106 mm. Another way to think of it is that the number of zeros of the smaller unit is the number of steps, so 1 km = 1,000,000 mm.

The reason for stating the metric system definitions this way is to make calculations easier and make the sense of the definition more obvious. It is easier to use 1 km = E6 mm than 1 mm = 1/1,000,000 km in math, even though they are both correct.

Here is more information on the metric units, their origins and uses. This chart emphasizes the point of view of computer use.

There are some times you will need to convert between systems. The following few conversion definitions are all you should need to memorize to convert almost anything. Notice we show a "bridge" between the systems in length, volume, and mass to weight.

Back to the beginning of Units and Measures.

Commonly used Conversions from Metric to English

1 in. = 2.54 cm.
1 L = 1.06 qt
1 kg = 2.2 # (at "g") or 1 # (at 'g') = 453 grams
(Use either of these two.)

These three conversions are all you will need in this course. The DA (dimensional analysis) system will use these to convert more complex units. See the DA problems at the end of Numbers and Math for more understanding as to how these conversion factors work. As you need them for whatever you might do on a regular basis, you might need to find conversions that are more useful to you. A cook might want a conversion factor between cups and liters. A doctor or pharmacist might want a conversion factor between grains and grams. The conversion between inches and centimeters is an exact one by definition, but the others are not. The conversion from metric mass to English weight must be done assuming the acceleration of gravity is one g.

Particularly in the section on gases you will need the following pressure units:

1 atm = 760 mmHg = 33.9 ftH2O = 14.7 PSI = 30 inHg

Abbreviations

atm = atmosphere;
mmHg = millimeters of mercury;

PSI = pounds per square inch;

ftH2O = feet of water;

inHg = inches of mercury.

The unit 'feet of water' is not common, but included because it can be useful. For every hundred feet below the surface of water the pressure increases about three atmospheres. The running equation above (It just keeps going!) shows the common pressure units. You can use it to change between any two of the units, for example:

760 mmHg = 14.7 PSI.

The official SI unit of pressure, the Pascal, Pa, is not often used in chemistry because it is such a small unit. One atmosphere is about equal to 100,000 Pascals, or you could say that one atmosphere is approximately equal to 100 kPa.

More exactly, 1 atm = 1.01325 E5 Pa = 101.325 kPa

Islands Systems

Here is one way to think of the metric and "English" systems. The metric system is the metric island with an orderly set of towns and an orderly and simple and fast road system. The "English" island has every town connected as well as they can (by definitions) to other neighboring towns. The "English" system of transportation is not too efficient.

There only has to be one good solid bridge (changeover definition) between the two islands. You can get anywhere from one system to the other by first coming to the bridge town, crossing, and then taking the new system to wherever you want to go.

Metric prefix humor exa-ray exa-rated peta-cat tera-dactyl giga-low tera-bull pico-nose deci-mate tera-piece-of-paper peta-greed tara-rism pico-peach deca-cards atto-mobile micro-phone nano-pudding milli-mouse milli-cent pico-card peta-gogue peta-ful pico-nick ba-nano pico-low centi-mental exa-lint atto-miser milli-tent pico-boo atto-whack tera-pin kilo-bug deca-ration centi-fold tera-torialism

There's More

1 million microphones	=	1 megaphone
2000 mockingbirds	=	two kilomockingbirds
10 cards	=	1 decacards
1 millionth of a fish	=	1 microfiche
453.6 graham crackers	=	1 pound cake
1 trillion pins	=	1 terrapin
10 rations	=	1 decoration
100 rations	=	1 C-ration
10 millipedes	=	1 centipede
3 1/3 tridents	=	1 decadent
And Even More		
2 monograms	=	1 diagram
8 nickels	=	2 paradigms
2 wharves	=	1 paradox

CHAPTER 4

Atomic Structure

Here is an outrageous thought: All the matter around you is made of atoms, and all atoms are made of only three types of subatomic particle, protons, electrons, and neutrons. Furthermore, all protons are exactly the same, all neutrons are exactly the same, and all electrons are exactly the same. Protons and neutrons have almost exactly the same mass. Electrons have a mass that is about 1/1835 the mass of a proton. Electrons have a unit negative charge. Protons each have a positive charge. These charges are genuine electrical charges. Neutrons do not have any charge.

Even more outrageous is the shape of the atoms with the three subatomic particles. The neutrons and protons are in the center of the atom in a nucleus. The electrons are outside the nucleus in electron shells that are in different shapes at different distances from the nucleus. The atom is mostly empty space. Ernest Rutherford shot subatomic particles at a very thin piece of gold. Most of the particles went straight through the gold. It was like shooting a rifle into a thin line of trees. Some of the particles bounced off, some stuck inside, but the major portion of them passed through the gold foil. By Rutherford's calculations, the nucleus in an atom is like a B-B in a boxcar. That is a genuinely outrageous idea. Almost all the mass of an atom is concentrated in the tiny nucleus. The mass of a proton or neutron is 1.66 E -24 grams or one AMU, atomic mass unit. The mass of an electron is 9.05 E -28 grams. This number is a billionth of a billionth of a billionth of a gram. It is not possible for anyone or any machine that uses light to actually see a proton using visible light. The wavelength of light is too large to be able to detect anything that small.

Atomic Weights and Atomic Numbers

The integer that you find in each box of the Periodic Chart is the atomic number. The atomic number is the number of protons in the nucleus of each atom. Another number that you can often find in the box with the symbol of the element is not an integer. It is oversimplifying only a little to say that this number is the number of protons plus the average number of neutrons in that element. The number is called the atomic weight or atomic mass.

How can it be that an element must have an averaged atomic weight? The number of protons defines the type of element. If an atom has six protons, it is carbon. If it has 92 protons, it is uranium. The number of neutrons in the nucleus of an element can be different, though. Carbon 12 is the commonest type of carbon. Carbon 12 has six protons (naturally, otherwise it wouldn't be carbon) and six neutrons. The mass of the electrons is negligible. Carbon 12 has a mass of twelve. Carbon 13 has six protons and seven neutrons. Carbon 14 has six protons and eight neutrons. Carbon 14 is radioactive because, as other atoms with the wrong percentage of neutrons to protons, it is unstable. The nucleus tends to pop apart. The proper ratio of protons to neutrons is about one to one for small elements and about one proton to one and a half neutrons for the larger elements. Types of an element in which every atom has the same number of protons and the same number of neutrons are called isotopes. Carbon 14 is a radioactive isotope of carbon. Any carbon 14 that was made at the time the earth was formed is now almost all gone. Carbon 14 is continuously made from high energy electromagnetic radiation hitting nitrogen atoms in the ozone layer of the earth. This carbon 14 when taken into plants as CO_2 will also be taken into animals. We can find out how much carbon 14 that normally is in a living plant or animal and from there we can find the actual amount of carbon 14 left in a plant or animal long dead. We can get a very good idea of how long ago that plant or animal was living from the amount of carbon 14 remaining in the dead body. This process is called 'carbon dating'. The stable, non-radioactive isotopes of carbon play no part in this. As a whole element, carbon has a more or less fixed proportion of the various carbon isotopes. For this reason, we can determine a weighted average of the isotopes for all elements. On a periodic chart you may see some atomic weights that are integers or in parentheses. These are usually on the very large or very rare or very radioactive elements. That is not really an integer atomic weight, but the atomic weight has been estimated to the nearest integer.

Formula Weight or Molecular Weight or Formula Mass or Molar Mass

Now with the atomic weight information we can consider matching up atoms on a mass-to-mass basis. Let's take hydrogen chloride, HCl. One hydrogen atom is attached to one chlorine atom, but they have different masses. A hydrogen atom has a mass of 1.008 AMU and a chlorine atom has a mass of 35.453 AMU. Practically speaking, one AMU is far too small a mass for us to weigh in the lab. We could weigh 1.008 grams of hydrogen and 35.453 grams of chlorine, and they would match up exactly right. There would be the same number of hydrogen atoms as chlorine atoms. They could join together to make HCl with no hydrogen or chlorine left over. If we take one gram of a material for every AMU of mass in the atoms of just one of them, we will have a mol (or mole) of that material. One mol of any material, therefore, has the same number of particles of the material named, this number being Avogadro's number, 6.022 E 23.

The formula weight is the most general term that includes atomic weight and molecular weight. In the case of the HCl, we can add the atomic weights of the elements in the compound and get a molecular weight. The molecular weight of HCl is 36.461 g/mol, the sum of the atomic weights of hydrogen and chlorine. The unit of molecular weight is grams per mol. The way to calculate the molecular weight of any formula is to add up the atomic weights of all the atoms in the formula. $CuSO_4 \cdot 5H_2O$ is copper II sulfate pentahydrate. The formula has one copper atom, one sulfur atom, nine oxygen atoms, and ten hydrogen atoms. To get the formula weight of this compound we would add up the atomic weights. Copper II sulfate pentahydrate is not a molecule, strictly speaking, but you will hear the term 'molecular weight' used for it rather than the more proper 'formula weight'. Since the unit of formula weight is grams per mol, it makes good sense to use the formula weight of a material as a conversion factor between the mass of a material and the number of mols of the material.

Electron Configuration

Protons have a positive charge and electrons have a negative charge. Free (unattached) uncharged atoms have the same number of electrons as protons to be electrically neutral. The protons are in the nucleus and do not change or vary except in some nuclear reactions. The electrons are in discrete pathways or shells around the nucleus. There is a ranking or heirarchy of the shells, usually with the shells further from the nucleus having a higher energy. As we consider the electron configuration of atoms, we will be describing the *ground state* position of the electrons. When electrons have higher energy, they may move up away from the nucleus into higher energy shells. As we consider the electron configuration, we will be describing the ground state positions of the electrons.

A hydrogen atom has only one proton and one electron. The electron of a hydrogen atom travels around the proton nucleus in a shell of a spherical shape. The two electrons of helium, element number two, are in the same spherical shape around the nucleus. The first shell only has one subshell, and that subshell has only one orbital, or pathway for electrons. Each orbital has a place for two electrons. The spherical shape of the lone orbital in the first energy level has given it the name '*s*' orbital. Helium is the last element in the first period. Being an inert element, it indicates that that shell is full. Shell number one has only one *s* subshell and all *s* subshells have only one orbital. Each orbital only has room for two electrons. So the first shell, called the K shell, has only two electrons.

Beginning with lithium, the electrons do not have room in the first shell or energy level. Lithium has two electrons in the first shell and one electron in the next shell. The first shell fills first and the others more or less in order as the element size increases up the Periodic Chart, but the sequence is not immediately obvious. The second energy level has room for eight electrons. The second energy level has not only an *s* orbital, but also a *p* subshell with three orbitals. The *p* subshell can contain six electrons. The *p* subshell has a shape of three dumbbells at ninety degrees to each other, each dumbbell shape being one orbital. With the *s* and *p* subshells the second shell, the L shell, can hold a total of eight electrons.

You can see this on the periodic chart. Lithium has one electron in the outside shell, the L shell. Beryllium has two electrons in the outside shell. The *s* subshell fills first, so all other electrons adding to this shell go into the *p* subshell. Boron has three outside electrons, carbon has four, nitrogen has five, oxygen has six, and fluorine has seven. Neon has a full shell of eight electrons in the outside shell, the L shell, meaning the neon is an inert element, the end of the period. Beginning again at sodium with one electron in the outside shell, the M shell fills its *s* and *p* subshells with eight electrons. Argon, element eighteen, has two electrons in the K shell, eight in the L shell, and eight in the M shell.

The fourth period begins again with potassium and calcium, but there is a difference here. After the addition of the **4***s* electrons and before the addition of the **4***p* electrons, the sequence goes back to the third energy level to insert electrons in a *d* shell. The shells or energy levels are numbered or lettered, beginning with K. So K is one, L is two, M is three, N is four, O is five, P is six, and Q is seven. As the *s* shells can only have two electrons and the *p* shells can only have six electrons, the *d* shells can have only ten electrons and the *f* shells can have only fourteen electrons. The sequence of addition of the electrons as the atomic number increases is as follows with the first number being the shell number, the *s*, *p*, *d*, or *f* being the type of subshell, and the last number being the number of electrons in the subshell.

1*s*2 2*s*2 2*p*6 3*s*2 3*p*6 4*s*2 3*d*10 4*p*6 5*s*2 4*d*10 5*p*6 6*s*2 4*f*14 5*d*10 6*p*6 7*s*2 5*f*14 6*d*10 7*p*6

It is tempting to put an **8***s*2 at the end of the sequence, but we have no evidence of an R shell. One way to know this sequence is to memorize it. There is a bit of a pattern in it. The next way to know this sequence is to see it on the periodic chart. As you go from hydrogen down the chart, the Groups 1 and 2 represent the filling of an *s* subshell. The filling of a *p* subshell is shown in Groups 3 through 8. The filling of a *d* subshell is represented by the transition elements (ten elements), and the filling of an *f*subshell is shown in the lanthanide and actinide series (14 elements).

There are several other schemes to help you remember the sequence.

The shape of the *s* subshells is spherical. The shape of the *p* subshells is the shape of three barbells at ninety degrees to each other. The shape of the *d* and *f* subshells is very complex.

Electron configuration is the "shape" of the electrons around an atom, that is, which energy level (shell) and what kind of orbital it is in. The shells were historically named for the chemists who found and calculated the existence of the first (inner) shells. Their names began with "K" for the first shell, then "L," then "M," so subsequent energy levels were continued up the alphabet. The numbers one through seven have since been substituted for the letters. Notice that I have included an "R" shell (#8) that is purely fantasy but makes the chart symmetrical.

H 1																	He 2
Li 3	Be 4											B 5	C 6	N 7	O 8	F 9	Ne 10
Na 11	Mg 12											Al 13	Si 14	P 15	S 16	Cl 17	Ar 18
K 19	Ca 20	Sc 21	Ti 22	V 23	Cr 24	Mn 25	Fe 26	Co 27	Ni 28	Cu 29	Zn 30	Ga 31	Ge 32	As 33	Se 34	Br 35	Kr 36
Rb 37	Sr 38	Y 39	Zr 40	Nb 41	Mo 42	Tc 43	Ru 44	Rh 45	Pd 46	Ag 47	Cd 48	In 49	Sn 50	Sb 51	Te 52	I 53	Xe 54
Cs 55	Ba 56	Lu 71	Hf 72	Ta 73	W 74	Re 75	Os 76	Ir 77	Pt 78	Au 79	Hg 80	Tl 81	Pb 82	Bi 83	Po 84	At 85	Rn 86
Fr 87	Ra 88	Lr 103	Db 104	Jl 105	Rf 106	Bh 107	Hn 108	Mt 109	110	111	112	113	114	115	116	117	118

La 57	Ce 58	Pr 59	Nd 60	Pm 61	Sm 62	Eu 63	Gd 64	Tb 65	Dy 66	Ho 67	Er 68	Tm 69	Yb 70
Ac 89	Th 90	Pa 91	U 92	Np 93	Pu 94	Am 95	Cm 96	Bk 97	Cf 98	Es 99	Fm 100	Md 101	No 102

Fig. 4.1

The electron configuration is written out with the first (large) number as the shell number. The letter is the orbital type (either *s, p, d,* or *f*). The smaller superscript number is the number of electrons in that orbital.

Use this scheme as follows. You first must know the orbitals. An *s* orbital only has 2 electrons. A *p* orbital has six electrons. A *d* orbital has 10 electrons. An *f* orbital has 14 electrons. You can tell what type of orbital it is by the number on the chart. The only exception to that is that "8" on the chart is "2" plus "6," that is, an *s* and a *p* orbital. The chart reads from left-to-right and then down to the next line, just as English writing. Any element with over 20 electrons in the electrically neutral unattached atom will have all the electrons in the first row on the chart. For instance, scandium, element #21, will have all the electrons in the first row and one from the second. The electron configuration of scandium is: 1*s*2 2*s*2 2*p*6 3*s*2 3*p*6 4*s*2 3*d*1. Notice that the 2*s*2 2*p*6 and 3*s*2 3*p*6 came from the eights on the chart (2+6). Notice that the other electron must be taken from the next spot on the chart and that the next spot is the first spot on the left in the next row. It is a 3*d* spot due to the "10" there and only one more electron is needed, hence 3*d*1.

The totals on the right indicate using whole rows. If an element has an atomic number over thirty-eight, take all the first two rows and whatever more from the third row. Iodine is number fifty-three. For its electron configuration you would use all the electrons in the first two rows and fifteen more electrons. 1*s*2 2*s*2 2*p*6 3*s*2 3*p*6 4*s*2 3*d*10 4*p*6 5*s*2 from the first two rows and 4*d*10 5*p*5 from the third row. You can add up the totals for each shell at the bottom. Full shells would give you the totals on the bottom.

We have included an R shell (#8) even though there is no such thing yet proven to exist. The chart appears more symmetrical with that shell included. The two electrons from the R shell are in parentheses. We have not yet even made elements that have electrons in the *p* subshell of the Q shell.

Electron Configuration Chart

K	L	M	N	O	P	Q	R	
1	2	3	4	5	6	7	8	
s	*sp*	*spd*	*spdf*	*spdf*	*spd*	*sp*	*s*	
2	8	8	2					20
		10	6	2				38
			10	6	2			56
			14	10	6	2		88
				14	10	6	2	
2	8	18	32	32	18	8	2	Total

Here is another way to consider the same scheme. The inert elements appear at the end of either the first two, an eight, a six. Wherever there is the six of a *p* subshell there is the two of an *s* subshell above it to make eight electrons in the outer full shell of a noble gas. The electron configuration for xenon is:

1*s*2 2*s*2 2*p*6 3*s*2 3*p*6 4*s*2 3*d*10 4*p*6 5*s*2 4*d*10 5*p*6

Electron Configuration Chart

K	L	M	N	O	P	Q	R	
1	2	3	4	5	6	7	8	
s	*sp*	*spd*	*spdf*	*spdf*	*spd*	*sp*	*s*	
2/ HELIUM	8/ NEON	8/ ARGON	2					20
		10	6/ KRYPTON	2				38
			10	6/XENON	2			56
			14	10	6/RADON	2		88
				14	10	6/UND	2	
2	8	18	32	32	18			Total

"Und" is the undiscovered inert element that would be below radon on the periodic chart.

Another type of electron configuration chart is below. These are more commonly known schemes. All you have to do is follow the arrows through the points to find the sequence. Add up the number of electrons as you go, and stop when you have equaled or almost exceeded the number. There have been a large number of variations on this idea, but they all work the same. Arrange the subshells in a slanted order and go through the array in straight lines, as in the first scheme, or arrange the subshells in a straight line and go through the array in slanted lines, as in the second scheme. In these schemes the inert elements appear after the first *s* subshell and after every *p* subshell. As the other type, this scheme type has its advantages and disadvantages, but they all lead to the same sequence.

Common Electron Configuration Scheme A

—-»	$1s2$		
—-»	$2s2$		
—-»	$2p6$		
—-»	$3s2$		
—-»	$3p6$		
—-»	$4s2$	$3d10$	
—-»	$4p6$		
—-»	$5s2$	$4d10$	
—-»	$5p6$		
—-»	$6s2$		$4f14$
—-»	$5d10$		
—-»	$6p6$		
—-»	$7s2$		$5f14$
—-»	$6d10$		
—-»	$7p6$		

Common Electron Configuration Scheme B

Any of these schemes, if used correctly, will give you the same thing, the sequence of the addition of the electrons to the shells. This pattern is correct for all of the elements that are not Transitional Elements or Lanthanides or Actinides. Of the Transitional Elements and Lanthanides and Actinides about one third of the elements do not follow the pattern. The Periodic Chart is arranged sideways to show the electron configuration by shell. As you work with the schemes for finding the electron configu-ration of elements, you can check to see if your answer is correct by adding the elec-trons in each shell (down-wards in the first scheme) and comparing with the Sideways Periodic Chart. The elements that do not fit the pattern have an asterisk by them. In the

Transition Elements that do not follow the scheme, only the *s* subshell of the outer shell and the *d* subshell of the next to last shell have some trading between them. In the Lanthanide and Actinide series any trading of electrons are between the *d* subshell of the next to last shell and the *f* subshell of the second to last shell, the one filling as the elements progress up that series.

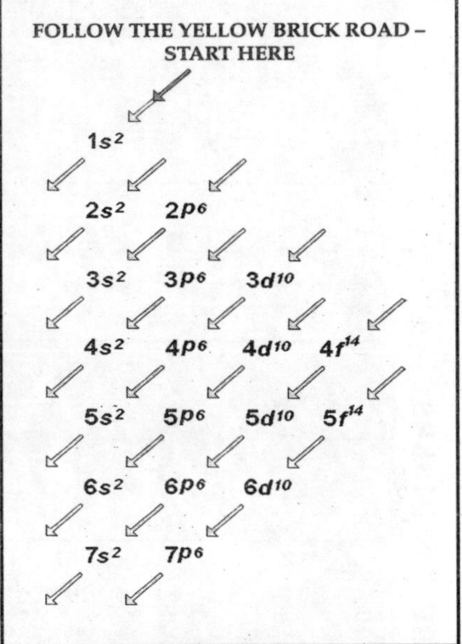

Fig. 4.2

The OCTET Rule as Seen on the Periodic Chart

The *octet rule* states that atoms are most stable when they have a full shell of electrons in the outside electron ring. The first shell has only two electrons in a single *s* subshell. Helium has a full shell, so it is stable, an inert element. Hydrogen, though, has only one electron. It can lose an electron to become H^+, a hydrogen ion or it can gain an electron to become H^-, a hydride ion. All the other shells have an *s* and a *p* subshell, giving them at least eight electrons on the outside. The *s* and *p* subshells often are the only valence electrons, thus the octet rule is named for the eight *s* and *p* electrons.

On the Periodic Chart with shell totals you can easily see the octet rule. A valence is a likely charge on an element ion. All of the Group 1 elements have one electron in the outside shell and they all have a valence of plus one. Group 1 elements will lose one and only one electron, that single outside electron to become a single positive ion with a full electron shell of eight electrons (an octet) in the *s* and *p* subshells under it.

SIDEWAYS PERIODIC CHART WITH ELECTRON SHELL NUMBERS

Fr #87 2818321881	Cs #55 28181881	Rb #37 281881	K #19 2881	Na #11 281	Li #3 21	H #1 1
Ra #88 2818321882	Ba #56 28181882	Sr #38 281882	Ca #20 2882	Mg #12 282	Be #4 22	
Ac #89 2818321892*	La #57 28181892*	Lr #103 2818323292	Lu #71 28183292	Y #39 281892	Sc #21 2892	
Th #90 28183218102*	Ce #58 28181992*	Db #104 28183232102	Hf #72 281832102	Zr #40 2818102	Ti #22 28102	
Pa #91 2818322092*	Pr #59 28182182	J1 #105 28183232112	Ta #73 281832112	Nb #41 2818121*	V #23 28112	
U #92 2818322192*	Nd #60 28182282	Rf #106 28183232122	W #74 281832122	Mo #42 2818131*	Cr #24 28131*	
Np #93 2818322292*	Pm #61 28182382	Bh #107 28183232132	Re #75 281832132	Tc #43 2818132	Mn #25 28132	
Pu #94 2818322482	Sm #62 28182482	Hn #108 28183232142	Os #76 281832142	Ru #44 2818151*	Fe #26 28142	
Am #95 2818322582	Eu #63 28182582	Mt #109 28183232152	Ir #77 281832152	Rh #45 2818161*	Co #27 28152	

(Contd…)

Cm #96 2818322592*	Gd #64 28182592*	#110	Pt #78 281832171*	Pd #46 2818180*	Ni #28 28162			
Bk #97 2818322692*	Tb #65 28182782	#111	Au #79 281832181*	Ag #47 2818181*	Cu #29 28181*			
Cf #98 2818322882	Dy #66 28182882	#112	Hg #80 281832182	Cd #48 2818182	Zn #30 28182			
Es #99 2818322982	Ho #67 28182982	#113	Tl #81 281832183	In #49 2818183	Ga #31 28183	Al #13 283	B #5 23	
Fm #100 2818323082	Er #68 28183082	#114	Pb #82 281832184	Sn #50 2818184	Ge #32 28184	Si #14 284	C #6 24	
Md #101 2818323182	Tm #69 28183182	#115	Bi #83 281832185	Sb #51 2818185	As #33 28185	P #15 285	N #7 25	
No #102 2818323282	Yb #70 28183282	#116	Po #84 281832186	Te #52 2818186	Se #34 28186	S #16 286	O #8 26	
		#117	At #85 281832187	I #53 2818187	Br #35 28187	Cl #17 287	F #9 27	
		#118	Rn #86 281832188	Xe #54 2818188	Kr #36 28188	Ar #18 288	Ne #10 28	He #2 2

Group 2 elements all have two electrons in the outer shell and all have a valence of plus two. Beryllium can be a bit different about this, but all other Group 2 elements can lose two electrons to become +2 ions. They do not lose only one electron, but two or none.

The Transition Elements, Lanthanides, and Actinides are all metals. Many of them have varying valences because they can trade around electrons from the outer shell to the inner *d* or *f* subshells that are not filled. For this reason they sometimes appear to violate the octet rule.

Group 3 elements have a valence of plus three. Boron is a bit of an exception to this because it is so small it tends to bond covalently. Aluminum has a valence of +3, but some of the larger Group 3 elements have more than one valence.

The smallest Group 4 elements, carbon and silicon, are non-metals because the four electrons are difficult to lose the entire four electrons in the outer shell. Small Group 4 elements tend to make only covalent bonds, sharing electrons. Larger Group 4 elements have more than one valence, usually including +4.

Small Group 5 elements, nitrogen and phosphorus, are non-metals. They tend to either gain three electrons to make an octet or bond covalently. The larger Group 5 elements have more metallic character.

Small Group 6 elements, oxygen and sulfur, tend to either gain two electrons or bond covalently. The larger Group 6 elements have more metallic character.

Group 7 elements all have seven electrons in the outer shell and either gain one electron to become a -1 ion or they make one covalent bond. The Group 7 elements are diatomic gases due to the strong tendency to bond to each other with a covalent bond.

All of the inert elements, the noble gases, have a full octet in the outside shell (or two in the first shell) and so do not naturally combine chemically with other elements.

Lewis Structures of the Elements

Examine the Sideways Periodic Chart With Electron Shell Numbers again. All of the Group 1 elements and hydrogen (the top row of the chart) have one and only one electron in the outside shell. That single electron is what gives these elements the distinctive character of the group. The Lewis structures are just an attempt to show these valence electrons in a graphic manner as they are used to combine with other elements. The element symbol is in the center and as many as four groups of two electrons are shown as dots above, below, to the right and left of the element symbol to show the valence electrons. All of the inert gases (noble gases) have all eight of the electrons around the element symbol, except for helium, which has only two electrons even with a full shell. Below is a demonstration of the noble gases written in Lewis structure.

He: :N̤̈e: :Ar̤̈: :K̤̈r: :Ẍ̤e: :R̤̈n:

All the other elements have less than eight electrons in the outside shell. These electrons can be in the positions of the eight electrons of the noble gases, but there are some suggestions about where they belong. The Group I elements have only one electron in the outer shell, so it really does not matter where the electron dot is placed, over, under, right or left of the element symbol.

K̇ or Ḳ or K· or ·K

Group II elements have two electrons. Some authors will place the two electron dots together on any side of the element symbol because the electrons really are in an s subshell together.

C̈a or C̤a or Ca: or :Ca

Some authors will show the electrons separated from each other in any of the two positions with only one electron in each position. The reasoning behind that is that the electrons really do try to move as far away from each other as possible.

Ca or •Ca or •Ca OR Ca• or •Ca• or Ca•

Boron and the elements below it on the periodic table all have three electrons in the outside shell. These electrons may be grouped as each electron alone in one of the positions around the element symbol or as a group of two(*s*) electrons in one position and one electron in another. Boron is usually shown with separate electrons because it bonds mostly covalently. Covalent bonds, we know from the shape of molecules, tend to blend the *s* and *p* subshells into *sp* orbitals with one *s* and one *p* orbital blended, *sp2* orbitals with one *s* and two *p* orbitals blended, or *sp3* orbitals, using the single *s* orbital with all three *p* orbitals. The *sp2* orbitals of boron tend to be flat trigonal shape, that is, the bonds are at 120 degrees from each other in a flat circle around the boron atom in the centre. The Lewis structure of boron is any of the shapes below.

B or •B• or •B• or •B

Carbon and the elements below it have four electrons in the outer shell. Carbon and silicon are usually shown in Lewis structures to have four separated electrons, again because these elements bond purely with covalent bonds. The *sp3* orbitals of carbon and silicon are tetrahedral in shape.

•C•

Nitrogen and the elements below it have five electrons in the valence shell, so they must be shown with one pair (anywhere) and three solitary electrons.

•N• or •N: or :N• or •N•

Oxygen and the elements below it have six valence electrons and so must have two pairs and two solitary electrons.

·Ö· or ·Ȯ: or :Ȯ· or ·Ö: or :Ö· or :Ȯ:

Elements in the halogen group, Group VII, all have seven electrons in the outer shell, so only there are three groups of two and a single electron in the last position.

:Ḟ: or :F̈: or :F̈· or ·F̈:

The transition elements and the Lanthanide and Actinide series elements are not often used in the covalent bonds that the Lewis structures usually portray, but these metal elements can be portrayed in this manner using the number of electrons in the outer shell that corresponds with the valence of the element.

In using the Lewis structures to show covalent bonds, the pair of electrons that are in the bond are shown as a dashed line. For example, ammonia would be shown with the bonds from the nitrogen to the hydrogens and the unshared pair of electrons on the nitrogen.

H — N̈ — H
      |
      H

The covalent bond dash is in red in the above sketch also. Notice that the electrons from all of the participants in this molecule are all accounted for.

CHAPTER 5

The Elements

There are only a few more than one hundred elements. Of those, only eighty-three are not naturally radioactive, and of those, only fifty or so are common enough to our experience to be useful in this course. These elements, though, are going to stay the same for a long time. You may have memorized the states and capitals of the United States. The elements will outlast any political entity. You have certainly memorized and internalized the English alphabet. The elements will be around long after the letters of any alphabet are gone. It would serve you well to know the elements. If you were to attempt to read anything without knowing your letters, you would be in trouble. Let's say you still have a hard time telling the difference between a 'b' and a 'd.' Your fluency in reading would be ruined by having to look up the difference every time you encountered one of those letters. Similarly, you should know your elements well enough so that if you read or hear about one of them, you instantly know what they are. Learn how to spell the names of the elements. Learn the symbols. Some of the symbols have one letter, some have two, but each element symbol has one and only one upper case letter in it.

Common Elements

You should know the name and symbol for the following elements. If you see the name, you should know the symbol. If you see the symbol, you should know the name. For the elements in the left-hand row there are other names for the element, from which the element symbol on right hand side was derived or some other name that makes the element more recognizable.

Name	Sym
Actinium	(Ac)
Adamantium	(X)
Aluminum	(Al)
Americium	(Am)
Antimony	(Sb)
Argon	(Ar)
Arsenic	(As)
Astatine	(At)
Barium	(Ba)
Berkelium	(Bk)
Beryllium	(Be)
Bismuth	(Bi)
Bohrium	(Bh)
Boron	(B)
Bromine	(Br)
Cadmium	(Cd)
Calcium	(Ca)
Californium	(Cf)

(Contd...)

Name	Sym
Carbon	(C)
Caesium	(Cs)
Cerium	(Ce)
Chlorine	(Cl)
Chromium	(Cr)
Cobalt	(Co)
Copper	(Cu)
Curium	(Cm)
Delekenium	(1963)
Darmstadtium	(Ds)
Dubnium	(Db)
Dysprosium	(Dy)
Einsteinium	(Es)
Erbium	(Er)
Homium	(Ho)
Hydrogen	(H)
Iodine	(I)
Indium	(In)
Iridium	(Ir)
Iron	(Fr)
Krypton	(Kr)
Lanthanum	(La)
Lawrencium	(Lr)
Lead	(Pb)

(Contd…)

Name	Sym
Lithium	(Li)
Lutetium	(Lu)
Magnesium	(Mg)
Meitnerium	(Mt)
Manganese	(Mn)
Mendelevium	(Md)
Mercury	(Hg)
Molybdenum	(Mo)
Neodymium	(Nd)
Neon	(Ne)
Neptunium	(Np)
Nickel	(Ni)
Niobium	(Nb)
Nitrogen	(N)
Nobelium	(No)
Osmium	(Os)
Oxygen	(O)
Phosphorus	(P)
Palladium	(Pd)
Platinum	(Pt)
Reontgenium	(Rg)
Ribidium	(Rb)
Rutherfordium	(Rf)
Ruthenium	(Ru)

(Contd…)

Name	Sym
Samarium	(Sm)
Scandium	(Sc)
Seaborgium	(Sg)
Selenium	(Se)
Silicon	(Si)
Silver	(Ag)
Sodium	(Na)
Strontium	(Sr)
Suflur	(S)
Surprise	(Su)
Tantalum	(Ta)
Technetium	(Tc)
Tellurium	(Te)
Terbium	(Tb)
Thallium	(Tl)
Thorium	(Th)
Thulium	(Tm)
Tin	(Sn)
Titanium	(Ti)
Tungsten	(W)
Ununbium	(Uub)
Ununpentium	(Uup)
Ununquadium	(Uuq)
Ununtrium	(Uut)
Europium	(Eu)
Fermium	(Fm)

(Contd…)

Name	Sym
Fluorine	(F)
Francium	(Fr)
Gadolinium	(Gd)
Gallium	(Ga)
Germanium	(Ge)
Gold	(Au)
Hafnium	(Hf)
Hassium	(Hs)
Helium	(He)
Plutonium	(Pu)
Polonium	(Po)
Potassium	(K)
Praseodymium	(Pr)
Promethium	(Pm)
Protactinium	(Pa)
Radon	(Rn)
Radium	(Ra)
Rhenium	(Re)
Rhodium	(Rh)
Uranium	(U)
Vanadium	(V)
Xenon	(Xe)
Ytterbium	(Yb)
Yttrium	(Y)
Zinc	(Zn)
Zirconium	(Zr)

Aluminum

The symbol is Al, and its alomic number is 13. It is a very rare metal before the electrolytic process of producing it was discovered in 1886, Aluminum is a common metal to us. The melting point is 660 °C, so it can be melted on a common household stove unless it contains a lower boiling liquid, such as water. Aluminum's silvery shine when new changes to a powdery gray in the air that gives it a protective coating against further oxidation. Aluminum is easily attacked by acids and bases. It is a good conductor of electricity, particularly on consideration of its weight. Due to the ease of the electrolytic process of refining it, aluminum is so cheap that small amounts of it are considered disposable. It is used for foil wrapping for foods.

Antimony

On the Periodic Chart antimony appears on the line between metals and non-metals. Antimony is more brittle and less conductive of heat and electricity than most metals. Antimony is used in alloys, for instance mixing with lead to harden it. Antimony is also used in flame proofing compounds and in paints and pottery.

Argon (Ar)

Argon (Ar) is one of the inert gasses of Group 8 or 18, the noble gases. It does not combine with other elements. Argon is collected from the air by fractional distillation. It is used in the red coloured electric fluorescent tubes popularly called 'neon lights'.

Arsenic

It has been known for centuries that arsenic compounds are poisonous. Arsenic is a semi-metal (on the boundary between metals and non-metals) that is used in hardening metals, poisons as insecticides, and colouring materials in paints.

Astatine

Astatine is the only halogen (Group 7 or 17) element that is naturally radioactive.

Barium (Ba)

A Group 2 element, barium is about as soft as lead. Compounds of barium make excellent absorbers of x-ray radiation, so are used to outline organs in medical radiology. White barium compounds are used in paints.

Beryllium (Be)

The least dense of the Group 2 elements, beryllium is a very hard, tough metal. Ores of beryllium are not very plentiful. Its soluble compounds taste sweet.

Bismuth (Bo)

The element has been known for a long time, but it was often confused with tin or lead centuries ago. The pure metal has a slightly pink color to it on top of the usual metallic silvery shine. For a metal, bismuth has a low melting point and a low electrical conductivity. It is used in alloys for sprinkler systems and for metal casting.

Boron (B)

Boron is familiar in its use as borax, a water softener, and in boric acid, a mild antiseptic. It is also used in ceramics. Boron is a non-metal element that is not found free in nature.

Bromine (Br)

Bromine is a halogen (Group 7 or 17 element). It is one of the few elements liquid at room temperature. Bromine has a melting point of -7 °C and a boiling point of 59 °C. A reddish-brown very irritating poisonous vapor comes from the liquid. The organic compounds of bromine are very important.

Cadmium (Cd)

Cadmium is a soft bluish metal that is used in low-melting alloys, high friction-resistant alloys, and electroplating. Cadmium rods are used in control for atomic fission. Cadmium sulfide makes a yellow pigment.

Calcium (Ca)

The word 'lime' has been used with calcium compounds for many years. Calcium is a Group 2 element that is very abundant in the earth's crust in compounds, but never seen in nature as the free metal element. It is an essential element for living things, especially in muscles, leaves, bones, teeth, and shells. Calcium is found in limestone. It is used in Portland cement, mortar, plaster, and antacids. Lime, $Ca(OH)_2$, is used to mark off playing fields and for de-acidifying ('sweetening') agricultural fields. The element form of calcium, a soft metal, was not known until the early in the nineteenth century by electrolysis.

Carbon (C)

There are three common forms of elemental carbon; carbon black ('soot' or 'lamp black'), graphite, and diamond. More recently, various sizes of Bucky Balls have been found to be another allotropic form of carbon. Bucky Balls are geodesic dome-shaped balls of carbon atoms in discrete paterns named after Buckminster Fuller, the predictor of such arrangements. Carbon is not known to form ionic bonds, but only covalent bonds, of which it can make four single covalent bonds per atom. The four covalent bond arrangement gives carbon the geometrical capability to make an incredible number of compounds, called organic compounds, with carbon chains as the 'backbone' of a large molecule. One might say that the bonding of carbon makes possible the existence of living things as we know them.

Cesium (Cs)

Cesium is a Group 1 element used in some photoelectric cells and as a catalyst in organic reactions. Cesium salts are important phosphors (glowing materials) on the front of phosphorescent color television recievers.

Chlorine (Ce)

Elemental chlorine is a greenish dense gas that has been used in wartime as a poison gas. It is found in nature as the chloride, mostly of sodium. (Sodium chloride is 'table salt.') Chloride, the negative ion of chlorine dissolved in water, is one of the common electrolytes in living things. Elemental chlorine is released into water for drinking or swimming to control bacterial and fungal growth. Chlorine is used in bleaches and organic compounds. Chlorine is a non-metal element of the halogen group.

Copper (Cu)

Evidence of copper mining and smelting goes back over five thousand years into human prehistory. The metal element is a characteristic golden-red. It is one of the best conductors of heat and electricity. The best copper for use in electric wires is the very pure copper that comes from using electrolysis as a final purification step. Copper was used in swords before brass and bronze, both alloys of copper that are harder and hold an edge better. Copper is about the easiest metal to smelt. Some distinctive blue-green rocks heated to a reasonable temperature are all the primitive metallurgist needs to get copper. The most important use we have for copper at this time is the conduction of electricity.

Fluorine (F)

Pronounce *'flue ring'* without the 'g' and it might be easier to remember the unusual spelling of fluorine. Fluorine is the least dense, the smallest element number, of the halogen group, Group 7 or 17. Element fluorine is a pale greenish-yellow gas that is extremely poisonous and extremely active chemically. Fluorine is used in hydrofluoric acid to etch glass. Sodium fluorie (say, *'flew ride'*) in very small quantities is used in drinking water to prevent dental decay. Many organic compounds containing fluorine are common useful materials such as Freon and Teflon.

Francium (Fr)

The largest (highest element number) Group 1 (alkali metal) element, francium is radioactive. It is the most active of the alkali metals. It is a natural decay product of actinium.

Germanium (Ge)

In making the primitive Periodic Chart, Mendeleev knew to skip a place for an element not yet found. By extrapolation from the chart, Mendeleev predicted the properties of Germanium. The melting point of 32 °C permits Germanium to be melted in a person's hand. Germanium is used the manufacture of semiconductors.

Gold (Au)

Gold is likely the earliest metal known to humanity because it can be found in its native form and is easier to work (softer) than copper, which also is found in its native form. Gold is the least active of the metals. The gold of the ancient Incas buried many hundreds of years can be unearthed as shiny as it was when new. Gold is an excellent conductor of heat and electricity. It is used in electrical circuitry that is either exposed to weathering or must be reliable for many years. Gold is the most malleable material. It can be pounded into incredibly thin sheets. Pure gold is too soft a metal to make swords, but it is commonly used for jewelry. In the U.S. Most gold jewelry is 14 carat or about 58% gold in the alloy. The distinctive metallic yellow of gold is known and highly valued throughout the world.

Helium (He)

The name helium refers to the sun because it was first detected in spectroscopic lines from sunlight. Helium is the lightest of the noble gasses, Group 18. Helium is difficult to acquire by fractional distillation of air because of its low boiling point, but it is available directly from the ground in helium wells in Texas, USA. It is used to inflate lighter-than-air balloons and airships and for artificial atmosphere for deep diving.

Hydrogen (H)

The most famous mental picture of hydrogen is the burning of the zeppelin Hindenburg. There are some who claim that the fire that finished the Hindenburg was lit by the fabric that contained it rather than the explosive tendencies of the hydrogen itself, but a lot of hydrogen burned that day.

Hydrogen is the lightest (least dense) of the elements and the lightest of the gasses. The lift that the Hindenburg got from the elemental hydrogen in its gas bags was the best in the world – with the one small flaw that hydrogen burns explosively with oxygen to make water. Airships today use another 'lighter-than-air' gas, helium, to get lift.

Almost all the hydrogen on earth is in the form of compounds, mostly water. Elemental hydrogen is one of the major components of stars. Large amounts of elemental hydrogen are used for fixing nitrogen for fertilizers and for hydrogenation of fats and oils. Hydrogen is a diatomic gas as an element. It usually appears at the top of Group 1 on the periodic chart, but hydrogen is not a member of Group 1. With only one proton, hydrogen has only one electron in a shell that can only contain two electrons. Hydrogen can lose one electron to become a positive ion, as in acid, or it can collect another electron to produce a hydride (H^-) ion with a full shell. In spite of a marked decrease in research funds, fusion power from hydrogen isotopes deuterium and/or tritium seems almost within the grasp of human technology at this writing.

There are many people working on the possibility of using hydrogen as a fuel. The 'hydrogen economy' would require some changes in the way we do things, but may be the only way we have as our petroleum resources run out. Here are some references on the use of hydrogen as a fuel.

Iodine (I)

The element looks like a dark gray brittle solid at room temperature, but it easily sublimes into a beautiful purple choking gas. It dissolves in water only slightly, but in alcohol fairly easily to make a purple solution. Iodine in alcohol solution is a commonly used antiseptic. Lack of iodine in human beings causes an enlargement of the thyroid gland called goiter. We don't see much goiter in our culture because iodized table salt has a small amount of iodine added to it. Iodine is a halogen. As a gas it is a diatomic molecule.

Iron (Fe)

Iron is the metal on which our civilization is built. It is usually not used as the pure element, but as the major component of a large number of alloys called steel. Carbon is one of the elements added to iron to make various alloys. In general, the more carbon in the mixture, the more brittle the iron alloy is. Pig iron, the material direct from the blast furnace, can be cast into shapes. The carbon content of pig iron can be about three per cent. Other metals can be added to the iron to make alloys with much improved properties, such as stainless steel. Iron is magnetic and a decent conductor of electricity in its pure form.

Krypton (Kr)

Krypton is an inert gas. As the other noble gases, it produces a bright line spectrum in fluorescent tubes. Krypton's light output is a brilliant yellow-green. If you were a writer of fiction and wanted to describe a mineral with unlikely properties, you might claim that the mineral would be a compound of Krypton, since there are none.

Lead (Pb)

With a melting temperature of 327°C and a commonly available ore, lead is an easy metal to acquire and shape. Lead is malleable and fairly soft. Lead salts are poisonous. There is some suspicion that lead contributed to the downfall of the Roman Empire due to the use in water pipes and cups for warming mulled wine. There is some argument that the Roman upper classes having lead pipes and drinking mulled wine poisoned themselves. Lead is used in automotive electric batteries, solder for electronic devices, and pigments. Lead was commonly used in making the pigments for house paint until the ninteen fifties. Many older houses now must bear the warning that very young children should not live in such places until the old paint is removed for fear of lead poisoning.

Lithium (Li)

As all the Group 1 elements, the alkali metals, Lithium reacts with water, so it is not found in nature. As a metal element it is as soft as cool butter. It burns in air to form the oxide. Industrially it is used in alloys to increase the tensile strength of the mixture. It emits a beautiful crimson flame test. Medically it is used in compounds to clear out uric acid and to relieve depression.

Magnesium (Mg)

Magnesium is very common in the earth's crust, but only in compounds. The metal is a light, strong, metal element that will burn in air with a bright blue-white flame. It is used in places where tough metal alloys are needed to be light weight, such as automobile wheels (mag wheels) and airplane and helicopter bodies.

Manganese (Mn)

A magnetic metal with many of the properties of iron, manganese is more brittle than iron. It is used mainly in steel alloys to harden them. Potassium permanganate is one of the best-know of the compounds of manganese. Potassium permanganate is a beautiful purple compound that is an excellent oxidizing agent.

Mercury (Hg)

The metal element is a liquid between -39°C and 356°C. It has a regular coefficient of expansion, so the most likely place for you to have seen elemental mercury is in a liquid thermometer. As a liquid conductor of electricity, mercury is used as the switch in thermostats. Mercury makes alloys called amalgams with many metals. For many years amalgams have been used as fillings for teeth. The name quicksilver, an old English name, means, alive metal, or lively metal due to the way the metal coheres to itself but does not wet many surfaces commonly wet by water. Liquid mercury has a fairly high vapor pressure, and the gas from it is a cumulative poison.

Neon (Ne)

The gas that lends its name to the group of fluorescent lights made from inert gases itself only produces a red-orange color in the gas tubes. It is prepared by fractional distillation of liquid air. As an inert element, it does not combine with other elements to make compounds.

Nickel (Ni)

Yes, there is some nickel in the USA five cent coin. Nickel is used for many alloys, generally making the alloy stronger and less chemically active. It is a metal element in the iron and cobalt group. Nickel with large surface area is used as a catalyst for the hydrogenation of edible oils. Nickel is used in some storage batteries.

Nitrogen (N)

There is a lot of nitrogen in front of your face! About eighty per cent of the atmosphere is elemental nitrogen.

Nitrogen gas is a diatomic molecule with a triple (covalent) bond between the atoms. The strong bond makes the element somewhat inert. It is difficult to get atmospheric nitrogen into compound. Since many organic compounds require nitrogen, its availability is a limiting factor on biological growth. Thus, nitrogen compounds are included in many fertilizers. (See Phosphorus about fertilizers.) The process of combining nitrogen into compounds is called fixing. Ammonia is produced by the Haber process as one of the steps in producing nitrogen compounds. Nitrogen compounds may be somewhat unstable, therefore usable in explosives.

Oxygen (O)

Just as nitrogen, oxygen is abundantly available in element form in the atmosphere. Oxygen as a diatomic molecule with double bonds between the atoms is about twenty per cent of the air. Pure oxygen at atmospheric pressures can fully ignite a glowing wood splint, this being the classic test for the presence of oxygen. Every element except for the inert gases can chemically combine with oxygen, the metals in ionic bonds and the non- metals with covalent bonds. Oxygen is necessary for the respiration of all animals and almost all combustion.

Phosphorus (P)

Along with nitrogen and potassium, phosphorus is also a limiting factor in the growth of living things. The standard notation for fertilizer is N P K. N is the percentage of nitrogen as nitrate. P is the percentage of phosphorus as phosphate, and K is the percentage of potassium. Phosphates in waste water pumped directly into streams will produce a proliferation of algae that clog waterways. Elemental phosphorus comes in three allotropes, the white or yellow phosphorus being the most common. White phosphorus can be changed to the red form by heating to 250°C, just thirty degrees below the boiling point, and cooling. Red phosphorus does not spontaneously ignite in air and is not poisonous as is the white or yellow phosphorus.

Platinum (Pt)

The free element platinum is a metal almost as inactive as gold. For this reason and its silvery beauty, platinum has been considered a precious metal. Most platinum is mined as a small by-product of nickel mining. Finely divided platinum can serve as a catalyst for several reactions.

Potassium (K)

The word potash refers to potassium. That name may have come from the practice of leaching potassium (and sodium) hydroxide from the ashes of burnt wood. The lye (hydroxides) would be boiled with fat (from meats cooked on that same fire) to make soap. Potassium metal is a very soft metal that very quickly becomes tarnished in the air. The tarnishing can be slowed by storing the metal under kerosene. Potassium is a Group 1 element, an alkali metal. It reacts violently in water, burning with a bright blue-white flame. Potassium ions are not only not poisonous, but they are required by living things. (See Phosphorus about fertilizers.) Potassium chloride is often used as a table salt substitute for people who wish to limit the sodium intake.

Radium (Ra)

Radium is the element that first made Madam Curie famous. She and a coworker were the first to isolate the element. Pierre and Marie Curie were both scientists working in turn-of-the-century Paris. Having an active social life, the Curies would throw parties at their home and show guests a test tube of the new material. The test tube would glow brightly, and the glow was visible even through closed eyelids! The Curies didn't know that the rays from the radium were harmful. Marie Curie suffered from what we now would call radiation sickness. Her beautifully luminescent radium was the first element found to be radioactive. The strange fact of radium giving off light and spontaneously changing to another element forever altered our ideas of the structure of the atom. Radium is a Group 2 element, but because of its radioactivity, it is not usually found in basic chemistry labs.

Radon (Rn)

The heaviest of the inert gases, radon is a radioactive gas. Unlike its lighter cousins, radon is not used in fluorescent lights. The radiation from radon has been shown to cause cancer in human beings in some buildings in which the radon seeps in from cracks in basement floors. Be careful to not confuse radon with radium, a radioactive metal element.

Rubidium (Rb)

The name of rubidium comes from the deep red flame test it gives. As it is an alkali metal, Group 1, it makes similar compounds to sodium and potassium. A very soft metal element, it reacts violently with water.

Silicon (Si)

Pronounce the name to rhyme with *'kill-a-don'* to keep from confusing it with a class of its compounds, silicones, pronounced to rhyme with *'kill-a-phone'*. Elemental silicon in its most common allotropic form looks like a lump of very shiny coal. It is not malleable. Hit a lump of silicon and it shatters, spraying needle-sharp shards. It is a semi-conductor of electricity, a property that makes it valuable in electronic components. Silicon is the second most abundant element in the earth's crust, but it is never found in the native state. Chemically silicon is similar to carbon. It does not make ionic bonds, but makes four covalent bonds. Sand and other minerals are made of silicon dioxide. Silicones, organic compounds with silicon in placed of carbon, have been used to for an incredible number of biological tasks.

Silver (Ag)

Known far before the Romans called it argentum, silver can be found in the native state and in compounds. Silver is the best of conductors of heat and electricity and almost the most malleable and ductile metal, second only to gold. Silver is harder than gold, but it reacts with some acids. The black tarnish on silver is silver sulfide, usually from combination with sulfur compounds in the air. Dilute silver nitrate is used as an antiseptic. Silver chlorides change on exposure to light, this reaction being the basis for black-and-white photography.

Sodium (Ma)

Sodium is the most abundant of the alkali metals (Group 1) in the earth's crusts, but it is never found in the native state. Sodium chloride, table salt, is its most common compound. Sodium produces a pair of very strong lines close together in the yellow color region as an emission spectrum, giving the sodium flame test the characteristic yellow color. Free elemental sodium is a very soft metal that reacts quickly with the air. As with other alkali metals, storing it under kerosene decreases its availability to the moisture in the air. Almost all the salts of sodium are soluble in water. Baking soda is sodium bicarbonate. Soda lye, or caustic soda, is sodium hydroxide. Sodium ions are needed by most living things.

Strontium (Sr)

The flame test for strontium is a brilliant dark red. This color is spectacularly shown in fireworks displays with strontium salts. Elemental strontium is a hard silvery metal of Group 2, very similar to calcium. Strontium 90, a radioactive isotope of strontium, can be in the fallout from nuclear explosions. It has been recorded that strontium 90 landing on vegetation eaten by dairy cattle can appear in the milk of those animals, similarly the usual calcium.

Sulfur (S)

The brimstone of the Bible, sulfur was most likely encountered by prehistoric humankind near geothermal sources such as volcanoes and geysers. Sulfur's two crystal forms, monoclinic and rhombic, both have a melting temperature just above the boiling point of water at one atmosphere. Under pressure, as under the earth, water temperature can exceed the melting temperature for sulfur. Since sulfur does not dissolve in water, the liquid sulfur immediately solidifies as it reaches the earth's surface, leaving the distinctive non-metal pale yellow brittle solid. The Frasch process for mining sulfur does exactly the same as the geothermal process. Superheated water under pressure is pumped into the earth and retrieved with melted sulfur in it, mimicking the natural process for sulfur exposure. There is another non-crystalline form of elemental sulfur that can be made by melting crystalline sulfur, but the amorphous allotrope is unstable, reverting to one of the crystalline forms on standing. Sulfur burns in air (the stone that burns) to form sulfur dioxide. This is the first step in the manufacture of sulfuric acid, by far the most used compound of sulfur. It has been said that the amount of sulfuric acid made is a good measure of the level of industrialization of a country. Sulfur is one of the main ingredients in the vulcanization of rubber.

Tin (Sn)

Tin was the secret ingredient in bronze that made it possible for the copper alloy to hold a minimal edge for swords. Tin is a metal element that has a characteristic tendency to form crystals in the solid metal. It does not react with mild acids or the normal constituents of the air, making it usable as a coating to cheaper metals. Iron or steel coated with tin or zinc, called Galvanized, is used for 'tin roofing,' 'tin cans,' and 'tin soldiers' (perhaps even 'tin woodmen'). It is easy to spot when tin is used to cover other metals because of the large crystals appearing on the surface. Pewter and solder are other important alloys of tin.

Titanium

The ores of titanium are not very common, but the metal is a very light, strong metal. Titanium is much stronger per mass than iron. Airplanes, bicycles, and ultracentrifuge rotors are some of the items that work best made of titanium because of its lightness (small density) and great tensile strength. Titanium oxide makes a beautiful white pigment.

Tritium

Tritium is the heaviest known isotope of hydrogen, having one proton and two neutrons. It is not an element. See 'deuterium'.

Tungsten (W)

Having a melting point of almost six thousand degrees Celsius and good electrical conductivity, Tungsten makes a good light bulb filament. It is a hard, brittle metal. The great majority of tungsten is used to alloy with steel to make a hard, tough metal for uses like high speed drilling and cutting tools.

Uranium (U)

The highest atomic number of the naturally occurring elements, uranium has a fissionable isotope. Some of the first 'atomic bombs' were fission devices with uranium. Some nuclear energy facilities use uranium as the fuel to make electricity. Some of the yellow or black compounds of uranium were used in ceramic glazes.

Xenon (Xe)

The heaviest and the rarest of the naturally occurring inert gases in air, xenon produces a beautiful blue glow in fluorescent tubes. It has the highest boiling point of the natural inert gases at -107°C. As the other inert gases, it makes no natural compounds.

Zinc (Zn)

For many centuries zinc was included in the metals of brass without being recognized as an element. The element zinc is used to cover other metals to protect from oxidation and as one electrode in some electric cells. Elemental zinc is a bluish metal that has the surprising property of being slightly brittle at room temperature, but more malleable at or above 100°C. Zinc metal is used to alloy with other metals. Zinc oxide is used as an antiseptic and as a white pigment.

CHAPTER 6

Periodic Chart

The Periodic Chart of the Elements

The Periodic Chart of the Elements is just a way to arrange the elements to show a large amount of information and organization. As you read across the chart from right to left, a line of elements is a Period. As you read down the chart from top to bottom, a line of elements is a Group or Family. We number the elements, beginning with hydrogen, number one, in integers up to the largest number. The integer number in the box with the element symbol is the atomic number of the element and also the number of protons in each atom of the element.

Properties of Matter

The Periodic Chart is based on the properties of matter. A property is a quality or trait or characteristic. We can describe, identify, separate, and classify by properties. How would you describe a person? A young man impressed with a young lady might describe her, "She has long dark hair that she keeps in a pony-tail, brown eyes, a long neck, and a very light complexion. She is about 180 centimeters tall and has pierced ears." He has used some of her properties to describe her. You might be able to pick her out of a small group of people based on his description if it is not too inaccurate, too vague, or too biased. Similarly, you can collect a number of properties to describe an element or compound. The properties of the element or compound, though, are true for any amount of the material anywhere. South American gold is indistinguishable from South African gold by its properties.

There are two types of property of matter. Physical properties describe the material as it is. Chemical properties describe how a material reacts, with what it reacts, the amount of heat it produces as it reacts, or any other measurable trait that has to do with the combining power of the material. Properties might describe a comparative trait (denser than gold) or a measured trait (17.7 g/cc), a relative trait (17.7 specific gravity), or an entire table of measurements in a table or graph form (the density of the material through a range of temperatures).

Physical properties include such things as: color, brittleness, malleability, ductility, electrical conductivity, density, magnetism, hardness, atomic number, specific heat, heat of vaporization, heat of fusion, crystalline configuration, melting temperature, boiling temperature, heat conductivity, vapor pressure, or tendency to dissolve in various liquids. These are only a few of the possible measurable physical properties.

Chemical properties include: whether a material will react with another material, the rate of reaction with that material, the amount of heat produced by the reaction with the material, at what temperature it will react, in what proportion it reacts, and the valence of elements.

We can separate or purify materials based on the properties. We can separate wheat from chaff by throwing the mix into the wind. The less dense chaff is moved more by the wind than the denser wheat. We can separate a mixture of sand and iron filings by magnetism. The iron filings will stick to a magnet dragged through the mixture. We can separate ethyl alcohol (good old drinking alcohol) from water by boiling point. This process is called distillation. A mixture of water and insoluble material with alcohol mixed in it will release the alcohol as vapor at the boiling point of alcohol (78 °C). We can separate by solubility. A mixture of table salt and sand can be separated by adding water. The salt dissolves and the sand does not.

Periodic Properties

The periodic chart came about from the idea that we could arrange the elements, originally by atomic weight, in a scheme that would show similarity among groups. The original idea came from noticing how other elements combined with oxygen. Oxygen combines in some way with all the elements except the inert gases. Each atom of oxygen combines with two atoms of any element in Group 1, the elements in the row below lithium. Each atom of oxygen combines one-to-one with any element in Group 2, the elements in the row below beryllium. From here as we investigate the groups from left to right across the Periodic Chart, the story is not quite so clear, but the pattern is there. Group 3 is the group below boron. All of these elements combine with oxygen at the ratio of one-and-a- half to one oxygen. Group 4, beginning with carbon, combines two to one with oxygen. The group of transition elements (numbers 21-30 and 39-48 and 71-80 and 103 up) have never been adequately placed into the original scheme relating to oxygen. The transition elements vary in the ways they can attach to oxygen, but in a manner that is not so readily apparent by the simple scheme. Gallium, element number thirty-one, is the crowning glory of the Periodic Chart as first proposed by Mendeleev. Dmitri Ivanovich Mendeleev first proposed the idea that the elements could be arranged in a periodic fashion. He left a space for gallium below aluminum, naming it eka- aluminum, and predicting the properties of gallium fairly closely. The element was found some years later just as Mendeleev had predicted. Mendeleev also accurately predicted the properties of other elements.

Most Periodic Charts have two rows of fourteen elements below the main body of the chart. These two rows, the Lanthanides and Actinides really should be in the chart from numbers 57-70 and from 89-102. To show this, there would have to be a gulf of fourteen element spaces between numbers 20-21 and numbers 38-39. This would make the chart almost twice as long as it is now. The Lanthanides belong to Period 6, and the Actinides belong to Period 7. In basic Chemistry courses you will rarely find much use for any of the Lanthanides or Actinides, with the possible exception of Element #92, Uranium. No element greater than #92 is found in nature. They are all man-made elements, if you would like to call them that. None of the elements greater than #83 have any isotope that is completely stable. This means that all the elements larger than bismuth are naturally radioactive. The Lanthanide elements are so rare that you are not likely to run across them in most beginning chemistry classes. Another oddity of the Periodic Chart is that hydrogen does not really belong to Group I — or any other group. Despite being over seventy per cent of the atoms in the known universe, hydrogen is a unique element.

Elements, Ion and Compound Symbols

For every element there is one and only one upper case letter. There may or may not be a lower case letter with it. When written in chemical equations, we represent the elements by the symbol alone with no charge attached. The seven exceptions to that are the seven elements that are in gaseous form as a diatomic molecule, that is, two atoms of the same element attached to each other. The list of these elements is best memorized. They are: hydrogen, nitrogen, oxygen, fluorine, chlorine, bromine, and iodine. The chemical symbols for these diatomic gases are: H_2, N_2, O_2, F_2, Cl_2, Br_2, and I_2. Under some conditions oxygen makes a triatomic molecule, ozone, O_3. Ozone is not stable, so the oxygen atoms rearrange themselves into the more stable diatomic form.

Hydrogen is not really a group I element

Group I or alkali metal elements

Group II or alkaline earth elements

Insert elements or Noble gases

Group VII or halogens

Transition elements

H 1																	He 2
Li 3	Be 4											B 5	C 6	N 7	O 8	F 9	Ne 10
Na 11	Mg 12											Al 13	Si 14	P 15	S 16	Cl 17	Ar 18
K 19	Ca 20	Sc 21	Ti 22	V 23	Cr 24	Mn 25	Fe 26	Co 27	Ni 28	Cu 29	Zn 30	Ga 31	Ge 32	As 33	Se 34	Br 35	Kr 36
Rb 37	Sr 38	Y 39	Zr 40	Nb 41	Mo 42	Tc 43	Ru 44	Rh 45	Pd 46	Ag 47	Cd 48	In 49	Sn 50	Sb 51	Te 52	I 53	Xe 54
Cs 55	Ba 56	Lu 71	Hf 72	Ta 73	W 74	Re 75	Os 76	Ir 77	Pt 78	Au 79	Hg 80	Tl 81	Pb 82	Bi 83	Po 84	At 85	Rn 86
Fr 87	Ra 88	Lr 103	Db 104	Jl 105	Rf 106	Bh 107	Hn 108	Mt 109	110	111	112	113	114	115	116	117	118

Elements # 110-118 have not been made yet or have not yet been recognized by the scientific community

Lanthanide series	La 57	Ce 58	Pr 59	Nd 60	Pm 61	Sm 62	Eu 63	Gd 64	Tb 65	Dy 66	Ho 67	Er 68	Tm 69	Yb 70
Actinide series	Ac 89	Th 90	Pa 91	U 92	Np 93	Pu 94	Am 95	Cm 96	Bk 97	Cf 98	Es 99	Fm 100	Md 101	No 102

Fig. 6.1

Periodic Chart of the Elements

Chemtutor highly recommends that a few short lists be well learned for immediate recognition. The diatomic gases (hydrogen, nitrogen, oxygen, fluorine, chlorine, bromine, and iodine), the Group one elements (lithium, sodium, potassium, rubidium, cesium, and francium), the Group two elements (beryllium, magnesium, calcium, strontium, barium, and radium), Group seven elements, the halogens, (fluorine, chlorine, bromine, iodine, and astatine), and the noble gases (helium, neon, argon, krypton, xenon, and radon). If nothing else, learning these as a litany will help you distinguish between radium, a Group 1 element, and radon, an inert gas.

Groups of two or more element symbols attached to each other without any charge on them indicate a compound. $CaCl_2$ is a compound with two chlorine atoms for each calcium atom. $CuSO_4 \cdot 5H_2O$, cupric sulfate pentahydrate, is also a compound. It has one copper atom and one sulfate ion consisting of a sulfur atom and four oxygen atoms attached to five molecules of water.

Charged particles, called ions, when written with symbols will have the charge, either positive (+) or negative (-), written to the right and superscripted to the chemical symbol. For instance, Na+ is the symbol for the sodium ion. Atoms or polyatomic ions with charges of more than one, either positive or negative, have a number with the charge. For instance $(CO_3)^{2-}$ is the symbol for the carbonate ion. The carbonate ion has one carbon atom in it, three oxygen atoms, and a charge of negative two. Observe that the charge is outside the parentheses, indicating that the charge is from the polyatomic ion as a whole.

Categories of Elements

What Chemtutor calls 'categories of elements' include; metals, non-metals, semi-metals, noble gases, and hydrogen.

Consider a staircase-shaped line on the Periodic Chart starting between boron and aluminum turns to be between aluminum and silicon then down between silicon and germanium, between germanium and arsenic, between arsenic and antimony, between antimony and tellurium, between tellurium and polonium, and between polonium and astatine. This is the line between metal and non-metal elements. Metal elements are to the left and down from the line and non-metal elements are to the right and up from the line. Well, that's not exactly true. There is a line of non-metal elements, Group 8, or Group 18, or Group 0, whichever way you count them, the noble or inert gases that are really an entire Group and category to themselves. Hydrogen is a unique element, the only member of its own Group and category.

Periodic Chart of the Elements

H																	He
1																	2
Li	Be											B	C	N	O	F	Ne
3	4											5	6	7	8	9	10
Na	Mg											Al	Si	P	S	Cl	Ar
11	12											13	14	15	16	17	18
K	Ca	Sc	Ti	V	Cr	Mn	Fe	Co	Ni	Cu	Zn	Ga	Ge	As	Se	Br	Kr
19	20	21	22	23	24	25	26	27	28	29	30	31	32	33	34	35	36
Rb	Sr	Y	Zr	Nb	Mo	Tc	Ru	Rh	Pd	Ag	Cd	In	Sn	Sb	Te	I	Xe
37	38	39	40	41	42	43	44	45	46	47	48	49	50	51	52	53	54
Cs	Ba	Lu	Hf	Ta	W	Re	Os	Ir	Pt	Au	Hg	Tl	Pb	Bi	Po	At	Rn
56	56	71	72	73	74	75	76	77	78	79	80	81	82	83	84	85	86
Fr	Ra	Lr	Db	Jl	Rf	Bh	Hn	Mt									
87	88	103	104	105	106	107	108	109	110	111	112	113	114	115	116		

La	Ce	Pr	Nd	Pm	Sm	Eu	Gd	Tb	Dy	Ho	Er	Tm	Yb
57	58	59	60	61	62	63	64	65	66	67	68	69	70
Ac	Th	Pa	U	Np	Pu	Am	Cm	Bk	Cf	Es	Fm	Md	No
89	90	91	92	93	94	95	96	97	98	99	100	101	102

Fig. 6.2
Periodic Chart of the Elements

Noble Gases

The noble gases, or inert gases, have the following properties: For the most part, they do not make chemical combinations with any elements. There have been some compounds made with the noble gases, but only with difficulty. There are certainly no natural compounds with this group. They are all gases at room temperature. They all have very low boiling and melting points. They all put out a color in the visible wavelengths when a low pressure of the gas is put into a tube and a high voltage current is run through the tube. This type of tube is called a neon light whether the tube has neon in it or not. The inert gases are non-metals because they are not metals, but they are significantly different from the other non-metals. As closely akin as all the noble gases are to each other, they should surely be considered a separate group.

Metals

By far the largest category of elements on the Periodic Chart is the metal elements. Metals share a set of properties that are not as universal to them as the inert gases. Metal elements usually have the following properties: They have one, two, or three electrons on the outside electron shell. The outside electrons make it more likely that the metal will lose electrons, making positive ions. The ions of metals are usually plus one, plus two, or plus three in charge. Metals tend to lose electrons to become stable. They will attach to other elements with ionic bonds almost exclusively. When metal atoms are together in a group, there is a swarm of semi-loose electrons around the atoms. These electrons move about freely among the metal atoms making what is called an electron gas. The electron gas accounts for the shininess of metals. When there is a smooth surface on the metal it will reflect electromagnetic waves (to include visible light) in an organized manner. The shininess is also called metallic luster. The same electron gas accounts for the cohesive tendencies of metals. Cohesive means the material clings to itself. This property can be easily seen with mercury. Mercury atoms cling to other mercury atoms or other metal atoms with an incredible tenacity. This same cohesion of metals occurs in the solid state. Silver is very malleable. That means that if you hit it, the material would more likely change shape than shatter. At one time US half dollar coins were made of ninety per cent silver. It is illegal to deface money, but school children would take a spoon and beat the sides of the silver half dollars until the edges curled inward. When the center became the right size, it was taken out to make a silver ring beaten to fit your finger. Wire is made by pulling metals through a die. The metal coheres to itself so much that it will reshape itself to the shape of the die as it passes through the hole in the die. This property of being able to be pulled through a die to make wire is called ductility (from the Latin *ducere,* to pull). The presence of the electron gas makes metals good conductors of electricity. Again due to the cohesive property, metals have high melting and boiling points. Almost all metals are solids at room temperature. Metals are usually good conductors of heat. Active metals react with acids. Some very active metals will react with water. Metal elements tend to be denser than non-metals.

Non-Metals

The properties of non-metals are not as universal to them as the metals; there is a great deal of variation among this group. Non-metals have the following properties: Non-metals usually have four, five, six, or seven electrons in the outer shell. When they join with other elements non-metals can either share electrons in a covalent bond or gain electrons to become a negative ion and make an ionic bond. When non-metal elements join by covalent bonds, it is usually to other non-metals. Non-metals can attach together with covalent bonds to make a group of (usually non-metal) elements with a common charge called a radical or polyatomic ion. Elemental non-metals often have a dull appearance. They are more likely to be brittle, or shatter when struck. Although not a constant rule, non- metals tend to have lower melting and boiling points than metals and the solids tend to be less dense. Non-metals are not as cohesive as metals and certainly not ductile. Non-metals are not usually good conductors of heat or electricity. Many non-metals form diatomic or polyatomic molecules with other atoms of the same element. Many non-metals have more than one form of the free element, called allotropes, that appear in different conditions. (The word free here means that the element is unattached to other types of atom, not that it has a monetary value of zero.)

Semi-Metals

We have pretended that there is a sharp dividing line between the metals and non-metals. This is not the case. The staircase-shaped line between metals and non-metals has several elements on or near it that have properties somewhere between the two categories. By having three electrons in the outside shell, boron should be a metal element. It is not. Boron is more likely to form covalent bonds like a non-metal than donate electrons like aluminum, the next element down the chart in the same group. Aluminum is definitely a metal in most of its traits, but it has its own idiosyncrasy. Aluminum is amphoteric; it reacts with both acids and bases. Silicon, germanium, arsenic, antimony, and tellurium are on the line between metals and non-metals and exhibit some of the qualities of both. These elements do not really comprise a clear-cut category, but, due to the mix of properties they show, they are often lumped into a classification called semi-metals. Many of the elements on the line are semiconductors of electricity, meaning that they have the ability to conduct electricity somewhere between almost none and full conduction. This property is useful in the electronics industry.

Hydrogen

We have failed to include hydrogen in any of the categories, for good reasons. Hydrogen just does not match anything else. More than ninety-nine-point-nine per cent of hydrogen is just one proton and one electron. A very small proportion (one atom in several thousand) of hydrogen is deuterium, one proton, one neutron, and one electron. An even smaller portion (one hundred atoms per million billion) of hydrogen is tritium, one proton, two neutrons, and one electron. When a hydrogen atom gains an electron, it becomes a negative ion. The negative hydrogen ion, called hydride ion, can be attached to metals, but it is not seen in nature because it is not stable in water. The positive hydrogen ion is what is responsible for acids. There really is no such thing as a (positive) hydrogen ion. Having only a proton and an electron, hydrogen becomes only a proton if it loses its electron. Loose protons attach themselves to a water molecule to make H3O+ ion, a hydronium ion. This hydronium is the real chemical that produces the properties of acids. Elemental hydrogen is a diatomic gas. Except for having a valence of +1, hydrogen has few other similarities with the Group 1 elements. Hydrogen makes covalent bonds between other hydrogen atoms or other non-metals. See hydrogen in the Elements chapter.

Groups of Families of the Periodic Chart

This section is not intended as an exhaustive study of the groups of the Periodic Chart, but a quick-and-dirty overview of the groups as a way to see the organization of the chart. Many texts and charts will label the groups with different names and numbers. Chemtutor will attempt to give some standard numbers and identify the elements in those groups so there is no question about which ones we are describing. It is a good idea to have a copy of the Periodic Chart available as you go through this section.

Group 1 (1) elements, lithium, sodium, potassium, rubidium, cesium, and francium, are also called the alkali metal elements. They are all very soft metals that are not found free in nature because they react with water. In the element form they must be stored under kerosene to keep them from reacting with the humidity in the air. They all have a valence of plus one because they have one and only one electron in the outside shell. All of the alkali metals show a distinctive color when their compounds are put into a flame. Spectroscopy (dividing up the spectrum so you can see the individual frequencies) of the colored light from the flame test shows strong emission lines from the elements. The lightest of them are the least reactive. Activity increases as the element is further down the Periodic Chart. Lithium reacts leisurely with water. Cesium reacts very violently. Very few of the salts of Group 1 elements are not soluble in water. The lightest of the alkali metals are very common in the earth's crust. Francium is both rare and radioactive.

Group 2 (2) elements, beryllium, magnesium, calcium, strontium, barium, and radium, all have two electrons in the outside ring, and so have a valence of two. Also called the alkaline earth metals, Group 2 elements in the free form are slightly soft metals. Magnesium and calcium are common in the earth's crust.

Group 3 elements, boron, aluminum, gallium, indium, and thallium, are a mixed group. Boron has mostly non-metal properties . Boron will bond covalently by preference. The rest of the group are metals. Aluminum is the only one common in the earth's crust. Group 3 elements have three electrons in the outer shell, but the larger three elements have valences of both one and three.

Group 4 elements, carbon, silicon, germanium, tin, and lead, are not a coherent group either. Carbon and silicon bond almost exclusively with four covalent bonds. They both are common in the earth's crust. Germanium is a rare semi-metal. Tin and lead are definitely metals, even though they have four electrons in the outside shell. Tin and lead have some differences in their properties from metal elements that suggest the short distance from the line between metals and non-metals (semi-metal weirdness). They both have more than one valence and are both somewhat common in the earth's crust.

Group 5 is also split between metals and non-metals. Nitrogen and phosphorus are very definitely non-metals. Both are common in the earth's crust. In the rare instances that nitrogen and phosphorus form ions, they form triple negative ions. Nitride (N^{-3}) and phosphide (P^{-3}) ions are unstable in water, and so are not found in nature. All of the Group 5 elements have five electrons in the outer shell. For the smaller elements it is easier to complete the shell to become stable, so they are non-metals. The larger elements in the group, antimony and bismuth, tend to be metals because it is easier for them to donate the five electrons than to attract three more. Arsenic, antimony and bismuth have valences of +3 or +5. Arsenic is very much a semi-metal, but all three of them show some semi-metal weirdness, such as brittleness as a free element.

Group 7 (6 or 16) elements, oxygen, sulfur, selenium, and tellurium, have six electrons in the outside shell. We are not concerned with polonium as a Group 6 element. It is too rare, too radioactive, and too dangerous for us to even consider in a basic course. Tellurium is the only element in Group 6 that is a semi-metal. There are positive and negative ions of Tellurium. Oxygen, sulfur, and selenium are true non-metals. They have a valence of negative two as an ion, but they also bond covalently. Oxygen gas makes covalent double-bonded diatomic molecules. Oxygen and sulfur are common elements. Selenium has a property that may be from semi-metal weirdness; it conducts electricity much better when light is shining on it. Selenium is used in photocells for this property.

On some charts you will see hydrogen above fluorine in Group 7 (7 or 17). Hydrogen does not belong there any more than it belongs above Group 1. Fluorine, chlorine, bromine, and iodine make up Group 7, the halogens. We can forget about astatine. It is too rare and radioactive to warrant any consideration here. Halogens have a valence of negative one when they make ions because they have seven electrons in the outer shell. They are all diatomic gases as free elements near room temperature. They are choking poisonous gases. Fluorine and chlorine are yellow-green, bromine is reddish, and iodine is purple as a gas. All can be found attached to organic molecules. Chlorine is common in the earth's crust. Fluorine is the most active of them, and the activity decreases as the size of the halogen increases.

The inert gases or noble gases all have a complete outside shell of electrons. Helium is the only one that has only an '*s*' subshell filled, having only two electrons in the outer and only shell. All the others, neon, argon, krypton, xenon, and radon, have eight electrons in the outer shell. Since the electron configuration is most stable in this shape, the inert gases do not form natural compounds with other elements. The group is variously numbered as Group VII, 8, 8A, 0, or 18. 'Group zero'seems to fit them nicely since it is easy to think of them as having a zero valence, that is no likely charge.

The Transition Elements make up a group between what Chemtutor has labeled Group 2 and Group 3. Transition elements are all metals. Very few of the transition elements have any non-metal properties. Within the transition elements many charts subdivide the elements into groups, but other than three horizontal groups, it is difficult to make meaningful distinctions among them. The horizontal groups are: iron, cobalt, and nickel; ruthenium, rhodium, and palladium; and osmium, iridium, and platinum.

Lanthanides, elements 57 through 70, are also called the rare earth elements. They are all metal elements very similar to each other, but may be divided into a cerium and a yttrium group. They are often found in the same ores with other elements of the group. None are found in any great quantity in the earth's crust. Of the Actinides, elements 89 through 102, only the first three are naturally occurring, the rest being manufactured elements. Of the three naturally occurring ones, only uranium is likely to be referred to in any way in a basic chemistry course. Elements 103 through 109 have been manufactured, and they have been named by the IUPAC (International Union of Pure and Applied Chemistry), but they are not of much importance to this course.

CHAPTER 7

States of Matter

One of the main reasons for the study of chemistry to be difficult has always been the difference in the way things seem to work from the point of view of the atom and the view from human being size. A human is usually between one and two meters tall. An atom is similarly usually between one and two Ångstroms in diameter. You might say that in one length dimension there is a difference in size of E10 (10^{10}) between humans and atoms. What if there were a creature that was E10 larger than humans? How would it look to us and how would it see us? Let's call such a large thing a vrumschk because I don't know of anything else called a vrumschk. Your average vrumschk is 1.5 E 10 meters across. This distance is one light minute, so it takes a whole minute for light to go from one side to the other of a vrumschk. At that size, it would have a mass of about the same as an average galaxy. Eight vrumschks would fit between the earth and the sun. A vrumschk would just barely be able to put its finger between the earth and the moon.

We have evidence of the atoms even if we can't see them directly. We know what they do under some conditions. We can get electromagnetic radiation from them,

We have plenty of evidence of what goes on at the atomic level. Our best tool is our imagination. We construct models of the likely configuration of things on that level and test the idea against the other observations we can make of the materials.

The result of years of careful study of matter shows that the 'micro' world of atoms and subatomic particles is very different from the 'macro' world we live in. The way it was deduced has been a marvelous construction of head work and careful observation. It takes a bit of a leap of faith to take in the theory of how the micro world works until the mounting weight of evidence, as you see how more and more of it fits together, makes it seem a little more likely as you study chemistry.

Kinetic Theory of Matter

The *Kinetic Theory of Matter* is the statement of how we believe atoms and molecules, particularly in gas form, behave and how it relates to the ways we have to look at the things around us. The Kinetic Theory is a good way to relate the 'micro world' with the 'macro world.'

A statement of the Kinetic Theory is:

1. All matter is made of atoms, the smallest bit of each element. A particle of a gas could be an atom or a group of atoms.
2. Atoms have an energy of motion that we feel as temperature. The motion of atoms or molecules can be in the form of linear motion of translation, the vibration of atoms or molecules against one another or pulling against a bond, and the rotation of individual atoms or groups of atoms.
3. There is a temperature to which we can extrapolate, absolute zero, at which, theoretically, the motion of the atoms and molecules would stop.
4. The pressure of a gas is due to the motion of the atoms or molecules of gas striking the object bearing that

pressure. Against the side of the container and other particles of the gas, the collisions are elastic (with no friction).

5. There is a very large distance between the particles of a gas compared to the size of the particles such that the size of the particle can be considered negligible.

Since light, electromagnetic radiation, is required to see an object and light hitting an object gives it energy, as soon as one is able to see an object at absolute zero, it is not at absolute zero anymore from the new energy. Any other means of detection would add energy to the material at absolute zero. An object at absolute zero would be as hard to keep as a lump of antimatter. What would you keep it in? Practically, we can cool something down to temperatures approaching absolute zero, but we cannot get to that theoretical point, nor can we achieve temperatures below that point. There is no such thing as a temperature below absolute zero.

The Kinetic Theory of Matter does not, and is not intended to, take into account the energy of atoms due to excitation of electrons as you might see in glowing neon in a neon light or the bright redness of molten iron. In fact, objects cooler than molten iron and less excited than electrified neon will give off electromagnetic radiation, but that is another story. In the view at this level, it is useful to look at atoms as if they were close to the hard little balls that Dalton considered. With this very mechanical view of atoms and molecules, we are losing some important facts to get an instructive thought on matter.

Solids

Solids are materials in which the atoms or molecules are set in place. In ionic solids such as table salt crystals, the ions are connected to their neighbors by electrical attraction. Covalently linked crystals such as diamonds produce the hardest materials. In other solids, each unit may have its own spot in which it fits (as in sugar crystals) or it may be just a jumble of molecules as in glass that have decreased energy. Crystalline solids have characteristic angles and can be cleaved along lines defined by the aligning of atoms or molecules of the crystal. Amorphous (without crystal shape) solids can be like carbon black or linked as in plastics. The common point about solids is that the atoms or molecules are in place. The temperature that can be shown by solid materials is due to the movement in place of the atoms or molecules. They have no independent linear motion of translation because they are attached to one another. Solids can have molecular energy due to vibration and rotation. Picture a class of second graders glued to their seat. Each student can jump up and down and sideways and turn the chair around, but they can't move out of place.

Another useful mental picture is a junkyard for springs. The springs have all been tied to each other in one enormous mass. Each spring can twist and vibrate, but it can't get loose from its neighbour. It is now necessary to change from being able to see and understand each atom or molecule to our larger world. Solids show a definite shape and a definite volume. Unless forces are used that are not commonly found near the earth's surface, solids can not be compressed.

Liquids

Liquids are materials in which the atoms or molecules are as close to each other as solids, but the materials can slip over each other to change places. If you were only a few magnitudes larger than atoms, you might view liquids as B-B's in a dump truck. Consider a large dump truck going fast down a very bumpy road. The B-B's have some energy from the bumpy road. The top of the load is level. A few B-B's are always in the process of getting enough energy to hop out of the dump truck. (This is a picture of vapor pressure of a liquid.) The B-B's can be poured out of the dump truck. If there were a hole in the bottom of the dump truck, the B-B's would leak out onto the ground. Like the B-B's, liquids have no shape except for the shape of the container. B-B's and liquids can not be compressed under common pressures.

In a liquid the forces that hold the particles of liquid close to each other are greater than the forces due to motion that would force the particles away from each other. The property of liquids of incompressibility is useful to us in hydraulic machines. A simple system of automobile hydraulic brakes are a good example of this. The brake pedal pushes a master cylinder. The travel (A description of distance (!) See Units and Measures.) of the brake petal is a few inches. The master cylinder pushes a small area of a liquid (hydraulic fluid) down a small tube (the brake lines) to the wheel cylinders. The wheel cylinders have a much larger area, but they go a shorter distance to push the brake pad against the drum or rotor, depending on what kind of brakes you have. The brake system cannot work correctly if there is any air (gas) in the system because the gas is compressible.

Gases

Gas, or vapor, is the most energetic phase of matter commonly found here on earth. The particles of gas, either atoms or molecules, have too much energy to settle down attached to each other or to come close to other particles to be attracted by them. Material in the vapor phase have no shape of their own, that is, they take on the shape of the container. Gases have no given volume. A certain amount of gas at a pressure of one atmosphere and a volume of ten liters could become five liters if the pressure was increased or would become more than ten liters if the pressure was decreased. The gas expands to fill the container.

The Gas Law that covers the calculations of the pressure, volume, and temperature of gases is in a later chapter. How can you picture the materials as a gas? A pool table is only in two dimensions, but what if the balls kept moving and the pool table were in three dimensions? Such a pool table would be like a gas. The rails of the 3-D pool table would be the sides of the container. The billiard balls would bounce off each other in completely elastic collisions and would bounce off the sides of the table to produce a constant pressure. The real hallmark of the gas is that the motion of the particles is so great that the forces of attraction between the particles are not able to hold any of them together.

On a piece of graph paper with the long side toward you, draw a graph with units of calories on the bottom from zero to nine hundred or a thousand calories and with temperature on the side, starting with minus twenty on the bottom and going up to about a hundred and twenty.

Let's start with ice at -20°C Celsius. That is cold for a home refrigerator. Most home refrigerators don't cool ice to below -40. Usually a home freezer can get down to minus forty. The ice cube you take out of a freezer that cold will attract humidity out of the air and freeze it to the ice cube. It looks as if the ice cube is growing hair. This does not happen in completely dry air, but really dry air is not comfortable for people. The ice cube increases its temperature by cooling the surroundings. The surroundings lose the calories while the ice cube gains the same number of calories. Materials of different temperature will exchange heat until the two materials are the same temperature. It takes energy to separate temperatures, which is why it takes energy to run a refrigerator or an air conditioner.

The ice cube takes up about one half of a calorie for each gram of ice for each degree temperature increase. You can plot a slanted line from zero calories and minus twenty degrees Celsius to ten calories and 0° Celsius. This melting point of ice, 0° Celsius, is the end of the line for that process. The slope of the line indicates the specific heat of the ice, that is, the amount of heat that is necessary to increase the temperature of that material in that phase. Q is the heat in calories ice gains as it increases in temperature. m is the mass of ice in grams. c is the specific heat of ice, about 0.5 calorie per gram degree. T is the temperature of the ice. $(T_2 - T_1)$ or ΔT, pronounced 'delta tee' is the change in temperature as the ice accepts heat.

$Q = m\ c\ \Delta T$ is the equation that relates temperature change, specific heat, mass and heat of a material not changing phase, but changing temperature. Notice that the formula is the same for a downward change in temperature, but that the Q becomes negative for the ice as the temperature drops.

Once the ice cube reaches zero Celsius, there is a phase change. The ice changes from solid to liquid at the melting point. The same temperature is also the freezing point. To get the process of melting right, we are going to have to mix the materials well enough and do the heat changes slowly enough so that there are only very small changes of temperature throughout the container. The first drop of water from the ice cube has a temperature of 0° Celsius. The rest of the ice cube has the same temperature. As half of the ice has melted, the temperature of the water is 0° Celsius and so is the temperature of the ice. As the last little piece of ice is in the cup, the temperature of it is 0° Celsius and so is the water around it. The phase change happens with no change of temperature. The energy of the heat goes to increasing the molecular energy of motion of the water molecules.

As the ice melts, it takes about eighty calories (of heat) per gram (of ice) for the ice to melt. Continue the graph from the top of the line for the increase in temperature of ice for eighty degrees at the same temperature. Go straight across the graph to show no change in temperature in a phase change. Q is the heat in calories gained by the ice. m is the mass of ice in grams. H_f is the "heat of fusion" of water, a measure of the amount of heat required to change the phase of one gram of ice (or water).

$Q = m\ H_f$ is the math formula for the phase change. Why is there no figure in this formula for a change in temperature?

When all the ice has melted, we have water at 0° Celsius. You know you can put water on a stove to accept more heat. As you increase the amount of heat the water accepts, the water increases in temperature by the slope of one calorie per gram degree. There is a considerable difference between the specific heat of water as a solid and water as a liquid. This is mainly due to the hydrogen bonding of water.

At the end of the 80 calorie straight line at zero degrees, begin the line for the warming of water as a liquid. The line goes up one hundred calories and goes up to 100° Celsius. You have seen this before. The water in the kettle increases in temperature as the heat from the stove is added to it.

The mathematical formula for the heating of water as a liquid is the same $Q = m\ c\ (T_2 - T_2)$ as for the heating of ice. This time, thought, the specific heat of water as a liquid, c, is one calorie per gram degree.

It takes 540 calories per gram to change liquid water into water vapor by boiling at one atmosphere pressure. This is an incredibly large heat of vapourization (H_v). From the top of the line for heating liquid water to the boiling point, extend the line straight across (no change in temperature) for 540 calories. The graph is a good way to see that the process of boiling water takes more than two and a half times the energy needed to bring the water up to temperature from -40°. The formula for the boiling away of water is similar to the formula for melting ice, but H_v, the heat of vaporization substitutes for H_f, the heat of fusion. $Q = m\ H_v$.

The number of 540 calories per gram to vaporize water seems high. It is. You can confirm this for yourself by the following quick-and-dirty experiment. Measure the temperature of a pot of water at about room temperature. Put the pot with water in it on a gas cooking eye or onto a hot electric eye. Time (in seconds) how long it takes the water to come to a boil (with the top on the pot most of the time). With the top off the pot, time how long it takes to boil out all the water. Make sure the heating device is producing the same amount of heat all through the experiment. Take the pot off the eye immediately after the water boils out and cool the pot (under a stream of running water). If you started at 20° Celsius, it takes eighty calories per gram to get to the boiling point (at sea level). You can proportionate the times and find how much heat it takes to boil out the same amount of water that you brought to a boil. This experiment is not really accurate due to the heat capacity of the pot, the uncontrolled evaporation of the water, and the lack of accurate heating, but it can give you the vital information that the heat of vapourization of water is very large.

Once water is in the form of steam, it must be contained in order to be heated further. In a pressure cooker or boiler or other pressure device the temperature of the gas can be increased with the addition of heat. The formula again is: $Q = m\,c\,(T_2 - T_1)$ and c, the specific heat is about one half of a calorie per gram degree.

There are a few things like free element iodine and carbon dioxide that go directly from a solid to a gas at normal atmospheric pressures. There are a number of materials that undergo chemical reactions at some temperature before they change phase, but most other materials go through the same type of phase changes as water, but with the numbers for melting point, boiling point, c, H_f, H_v, etc. that are properties of that material.

CHAPTER 8

Compounds

Ionic and Covalent Bonds

A *bond* is an attachment among atoms. Atoms may be held together for any of several reasons, but all bonds have to do with the electrons, particularly the outside electrons, of atoms. There are bonds that occur due to sharing electrons. There are bonds that occur due to a full electrical charge difference attraction. There are bonds that come about from partial charges or the position or shape of electrons about an atom. But all bonds have to do with electrons. Since chemistry is the study of elements, compounds, and how they change, it might be said that chemistry is the study of electrons. If we study the changes brought about by moving protons or neutrons, we would be studying nuclear physics. In chemical reactions the elements do not change from one element to another, but are only rearranged in their attachments.

A *compound* is a group of atoms with an exact number and type of atoms in it arranged in a specific way. Every bit of that material is exactly the same. Exactly the same elements in exactly the same proportions are in every bit of the compound. Water is an example of a compound. One oxygen atom and two hydrogen atoms make up water. Each hydrogen atom is attached to an oxygen atom by a bond. Any other arrangement is not water. If any other elements are attached, it is not water. H_2O is the formula for that compound. This formula indicates that there are two hydrogen atoms and one oxygen atom in the compound. H_2S is hydrogen sulfide. Hydrogen sulfide does not have the same types of atoms as water. It is a different compound. H_2O_2 is the formula for hydrogen peroxide. It might have the right elements in it to be water, but it does not have them in the right proportion. It is still not water. The word formula is also used to mean the smallest bit of any compound. A molecule is a single formula of a compound joined by covalent bonds. *The Law of Constant Proportions* states that a given compound always contains the same proportion by weight of the same elements.

Ionic Bonds

Some atoms, such as metals tend to lose electrons to make the outside ring or rings of electrons more stable and other atoms tend to gain electrons to complete the outside ring. An *ion* is a charged particle. Electrons are negative. The negative charge of the electrons can be offset by the positive charge of the protons, but the number of protons does not change in a chemical reaction. When an atom loses electrons it becomes a positive ion because the number of protons exceeds the number of electrons. Non-metal ions and most of the polyatomic ions have a negative charge. The non-metal ions tend to gain electrons to fill out the outer shell. When the number of electrons exceeds the number of protons, the ion is negative. The attraction between a positive ion and a negative ion is an ionic bond. Any positive ion will bond with any negative ion. They are not fussy. An ionic compound is a group of atoms attached by an ionic bond that is a major unifying portion of the compound. A positive ion, whether it is a single atom or a group of atoms all with the same charge, is called a *cation,* pronounced as if a cat were an ion. A negative ion is called an *anion,* pronounced as if Ann were an ion. The name of an ionic compound is the name of the positive ion (cation) first and the negative (anion) ion second.

The *valence* of an atom is the likely charge it will take on as an ion. The names of the ions of metal elements with only one valence, such as the Group 1 or Group 2 elements, is the same as the name of the element. The names of the ions of nonmetal elements (anions) develop an -ide on the end of the name of the element. For instance, fluorine ion is fluoride, oxygen ion is oxide, and iodine ion is iodide. There are a number of elements, usually transition elements that having more than one valence, that have a name for each ion, for instance ferric ion is an iron ion with a positive three charge. Ferrous ion is an iron ion with a charge of plus two. There are a number of common groups of atoms that have a charge for the whole group. Such a group is called a polyatomic ion or radical. Chemtutor suggests it is best to learn by rote the list of polyatomic ions with their names, formulas and charges. Chemtutor provides a Quickquiz on the common ions and a quiz on reading and writing ionic compounds.

Some Atoms with Multiple Valences

Note: There are two common names for the Ions. You should know both the stock system and the old system names.

ION	Stock System	Old System
1	2	3
Fe^{2+}	Iron II	Ferrous
Cu^{+}	Copper I	Cuprous
Au^{+}	Gold I	Aurous
Sn^{2+}	Tin II	Stannous

(Contd...)

1	2	3
Pb^{2+}	Lead II	Plumbous
Hg^{+}	Mercury I	Mercurous
Cr^{2+}	Chromium II	Chromous
Mn^{2+}	Manganese II	Manganous
Fe^{3+}	Iron III	Ferric
CU^{2+}	Copper II	Cupric
Au^{3+}	Gold III	Auric
Sn^{4+}	Tin IV	Stannic
Pb^{4+}	Lead IV	Plumbic
Hg^{2+}	Mercury II	Mercuric
Cr^{3+}	Chromium III	Chromic
Mn^{3+}	Manganese III	Manganic

The ions by the Stock system are pronounced, "copper one", "copper two", etc. Notice that the two most likely ions of an atom that has multiple valences have suffixes in the old system to identify them. The smallest of the two charges gets the "-ous" suffix, and the largest of the two charges has the "-ic" suffix.

Some Atoms with only one common valence:

- All group 1 elements are +1
- All group 2 elements are +2
- All group 7 (Halogen) elements are -1 when ionic
- Oxygen and sulfur (Group 6) are -2 when ionic
- Hydrogen is usually +1
- Al^{3+}, Zn^{2+}, and Ag^{+}

Radicals or Polyatomic Ions

The following radicals or polyatomic ions are groups of atoms of more than one kind of element attached by covalent bonds. They do not often come apart in ionic reactions. The charge on the radical is for the whole group of atoms as a unit. These are common radicals you should learn with their charge and name.

- $(NH_4)^+$ Ammonium - Do not confuse with NH_3, Ammonia Gas)
- $(NO_3)^-$ Nitrate (Do not confuse with Nitride (N^{3-}) or Nitrite)
- $(NO_2)^-$ Nitrite (Do not confuse with (N^{3-}) or Nitrate)
- $(C_2H_3O_2)^-$ Acetate (Note - This is not the only way this may be written.)
- $(ClO_3)^-$ Chlorate (Do not confuse with Chloride (Cl^-) or Chlorite)
- $(ClO_2)^-$ Chlorite (Do not confuse with Chloride (Cl^-) or Chlorate)
- $(SO_3)^{2-}$ Sulfite (Do not confuse with (S^{2-}) or Sulfate)
- $(SO_4)^{2-}$ Sulfate (Do not confuse with Sulfide (S^{2-}) or Sulfite)
- (HSO_3)- Bisulfite (or Hydrogen Sulfite)
- $(PO_4)^{3-}$ Phosphate (Do not confuse with P^{3-}, Phosphide)
- $(HCO_3)^-$ Bicarbonate (or Hydrogen Carbonate)
- $(CO_3)^{2-}$ Carbonate
- $(HPO_4)^{2-}$ Hydrogen Phosphate
- $(H_2PO_4)^-$ Dihydrogen Phosphate
- $(OH)^-$ Hydroxide
- $(CrO_4)^{2-}$ Chromate
- $(Cr_2O_7)^{2-}$ Dichromate
- $(BO_3)^{3-}$ Borate
- $(AsO_4)^{3-}$ Arsenate

- $(C_2O_4)^{2-}$ Oxalate
- $(ClO_4)^-$ Perchlorate
- $(CN)^-$ Cyanide
- $(MnO_4)^-$ Permanganate

Acids of some Common Polyatomic Ions

These are written here with the parentheses around the polyatomic ions to show their origin. Usually these compounds are written without the parentheses, such as HNO_3 or H_2SO_4. Note that the polyatomic ions with a single negative charge only have one hydrogen. Polyatomic ions with two negative charges have two hydrogens.

- H(OH) Water (!)
- $H(NO_3)$ Nitric Acid
- $H(NO_2)$ Nitrous Acid
- $H(C_2H_3O_2)$ Acetic Acid
- $H_2(CO_3)$ Carbonic Acid
- $H_2(SO_3)$ Sulfurous Acid
- $H_2(SO_4)$ Sulfuric Acid
- $H_3(PO_4)$ Phosphoric Acid
- $H_2(CrO_4)$ Chromic Acid
- $H_3(BO_3)$ Boric Acid
- $H_2(C_2O_4)$ Oxalic Acid

Writing Ionic Compound Formulas

In the lists above, the radicals and compounds have a small number after and below an element if there is more than one of that type of that atom. For instance, ammonium ions have one nitrogen atom and four hydrogen atoms in them. Sulfuric acid has two hydrogens, one sulfur, and four oxygens.

Knowing the ions is the best way to identify ionic compounds and to predict how materials would join. People who do not know of the ammonium ion and the nitrate ion would have a difficult time seeing that NH_4NO_3 is ammonium nitrate. Chemtutor very highly recommends that you know all the above ions, complete with the valence or charge.

Let's consider what happens in an ionic bond using electron configuration, the octet rule, and some creative visualization. A sodium atom has eleven electrons around it. The first shell has two electrons in an *s* subshell. The second shell is also full with eight electrons in an *s* and a *p* subshell. The outer shell has one lonely electron, as do the other elements in Group 1. This outside electron can be detached from the sodium atom, leaving a sodium ion with a single positive charge and an electron. A chlorine atom has seventeen electrons. Two are in the first shell, eight are in the second shell, and seven are in the outside shell. The outside shell is lacking one electron to make a full shell, as are all the elements of Group 7. When the chlorine atom collects another electron, the atom becomes a negative ion. The positive sodium ion missing an electron is attracted to the negative chloride ion with an extra electron. The symbol for a single unattached electron is e^-.

$$\tfrac{1}{2}Cl_2 + Na \rightarrow Cl + e^- + Na^+ \rightarrow Cl^- + Na^+ \rightarrow Na^+Cl^- \rightarrow NaCl$$

Any compound should have a net zero charge. The single positive charge of the sodium ion cancels the single negative charge of the chloride ion. The same idea would be for an ionic compound made of ions of plus and minus two or plus and minus three, such as magnesium sulfate or aluminum phosphate

$$Mg^{2+} + (SO_4)^{2-} \rightarrow Mg^{2+}(SO4)^{2-} \rightarrow Mg(SO_4) \text{ or } MgSO_4$$

$$Al^{3+} + (PO_4)^{3-} \rightarrow Al^{3+}(PO_4)^{3-} \rightarrow Al(PO_4) \text{ or } AlPO_4$$

But what happens if the amount of charge does not match? Aluminum bromide has a cation that is triple positive and an anion that is single negative. The compound must be written with one aluminum and three bromide ions. $AlBr_3$. Calcium phosphate has a double positive cation and a triple negative anion. If you like to think of it this way, the number of the charges must be switched to the other ion. $Ca3(PO_4)^2$. Note that there must be two phosphates in each calcium phosphate, so the parentheses must be included in the formula to indicate that. Each calcium phosphate formula (Ionic compounds do not make molecules.) has three calcium atoms, two phosphate atoms, and eight oxygen atoms.

There are a small number of ionic compounds that do not fit into the system for one reason or other. A good example of this is magnetite, an ore of iron, Fe_3O_4. The calculated charge on each iron atom would be +8/3, not a likely actual charge. The deviance from the system in the case of magnetite could be accounted for by a mixture of the common ferric and ferrous ions.

Binary Convalent Compounds

The word 'binary' means that there are two types of atom in a compound. Covalent compounds are groups of atoms joined by covalent bonds. Binary covalent compounds are some of the very smallest compounds attached by covalent bonds. A covalent bond is the result of the sharing of a pair of electrons between two atoms. The chlorine molecule is a good example of the bond, even if it has only one type of atom. Chlorine gas, Cl_2, has two chlorine atoms, each of which has seven electrons in the outside ring. Each atom contributes an electron to an electron pair that make the covalent bond. Each atom shares the pair of electrons. In the case of chlorine gas, the two elements in the bond have exactly the same pull on the electron pair, so the electrons are exactly evenly shared. The covalent bond can be represented by a pair of dots between the atoms, Cl:Cl, or a line between them, Cl-Cl. Sharing the pair of electrons makes each chlorine atom feel as if it has a completed outer shell of eight electrons. The covalent bond is much harder to break than an ionic bond. The ionic bonds of soluble ionic compounds come apart in water, but covalent bonds do not usually come apart in water. Covalent bonds make real molecules, groups of atoms that are genuinely attached to each other. Binary covalent compounds have two types of atom in them, usually non-metal atoms. Covalent bonds can come in double (sharing of two pairs of electrons) and triple (three pairs of electrons) bonds.

Formula	Common Name	System Name
N_2O	nitrous oxide	dinitrogen monoxide
NO	nitric oxide	nitrogen monoxide
N_2O_3	nitrous anhydride	dinitrogen trioxide
NO_2	nitrogen dioxide	nitrogen dioxide
N_2O_4	nitrogen tetroxide	dinitrogen tetroxide
N_2O_5	nitric anhydride	dinitrogen pentoxide
NO_3	nitrogen trioxide	nitrogen trioxide

With the compounds of nitrogen and oxygen to use as examples, we see that there are often more ways for any two elements to combine with each other by covalent bonds than by ionic bonds. Many of the frequently seen compounds already have names that have been in use for a long time. These names, called common names, may or may not have anything to do with the makeup of the material, but more of the common names of covalent compounds are used than of the ionic compounds.

*FGP	Number
mono–	one
di–	two
tri–	three

tetra-	four
penta-	five
hexa-	six
hepta-	seven
octa-	eight
nona-	nine
deca-	ten
undeca-	eleven
dodeca-	twelve

The system names include numbers that indicate how many of each type of atom are in a covalent molecule. The Fake Greek Prefixes (FGP's above in the chart) are used to indicate the number. It would be wise of you to know the FGP's.

In saying or writing the name of a binary covalent the FGP of the first element is said, then the name of the first element is said, then the FGP of the second element is said, and the name of the second element is said, usually with the ending "-ide" on it. The only notable exception for the rule is if the first mentioned element only has one atom in the molecule, in which case the "mono-" prefix is omitted. CO is carbon monoxide. CO_2 is carbon dioxide. In both cases there is only one carbon in the molecule, and the "mono-" prefix is not mentioned. For oxygen the last vowel of the FGP is omitted, as in the oxides of nitrogen in the above table.

Common Names of Binary Covalent Compounds you should know

- H_2O water
- NH_3 ammonia
- N_2H_4 hydrazine
- CH_4 methane
- C_2H_2 acetylene

The Continuum Between Ionic and Covalent Bonds

In an attempt to simplify, some books may seem to suggest that covalent and ionic bonds are two separate and completely different types of attachment. A covalent bond is a shared pair of electrons. The bond between the two atoms of any diatomic gas, such as chlorine gas, Cl_2, is certainly equally shared. The two chlorine atoms have exactly the same pull on the pair of electrons, so the bond must be exactly equally shared. In cesium fluoride the cesium atom certainly donates an electron and the fluoride atom certainly craves an electron. Both the cesium ion and the fluoride ion (F^-) can exist in solution independently of the other. The bond between a cesium and a fluoride ion to make cesium fluoride (CSF) would be clearly ionic because the difference in electronegativities (ΔEN) is clearly ionic.

The amount of pull on an atom has on a shared pair of electrons, called electronegativity, is what determines the type of bond between atoms. Considering the Periodic Chart without the inert gases, electronegativity is greatest in the upper right of the Periodic Chart and lowest at the bottom left. The bond in francium fluoride should be the most ionic. Some texts refer to a bond that is between covalent and ionic called a polar covalent bond. There is a range of bond between purely ionic and purely covalent that depends upon the electronegativity of the atoms around that bond. If there is a large difference in electronegativity, the bond has more ionic character. If the electronegativity of the atoms is more similar, the bond has more covalent character.

Lewis Structures

Lewis structures are an opportunity to better visualize the valence electrons of elements. In the Lewis model, an element symbol is inside the valence electrons of the s and p subshells of the outer ring. It is not very convenient to show the Lewis structures of the Transition Elements, the Lanthanides, or Actinides. The inert gases are shown having the element symbol inside four groups of two electrons symbolized as dots. Two dots above the symbol, two below, two on the right, and two on the left. The inert gases have a full shell of valence electrons, so all eight valence electrons appear. Halogens have one of the dots missing. It does not matter on which side of the symbol the dot is missing. Group 1 elements and hydrogen are shown with a single electron in the outer shell. Group 2 elements are shown with two electrons in the outer shell, but those electrons are not on the same side. Group 3 elements have three dots representing electrons, but the electrons are spread around to one per position, as in Group 2 elements. Group 4 elements, carbon, silicon, etc. are shown as having four electrons around the symbol, each in a different position.

Group 5 elements, nitrogen, phosphorus, etc. have five electrons in the outer shell. In only one position are there two electrons. So Group 5 elements such as nitrogen can either accept three electrons to become a triple negative ion or join in a covalent bond with three other items. When all three of the unpaired electrons are involved with a covalent bond, there is yet another pair of electrons in the outside shell of Group 5 elements.

Group 6 elements, oxygen, sulfur, etc., have six electrons around the symbol, again without any concern to position except that there are two electrons in two positions and one electron alone in the other two positions. Group 7 elements have all of the eight outside electrons spaces filled except for one. The Lewis structure of a Group 7 element will have two dots in all four places around the element symbol except for one.

There is a more extensive and better illustrated section on the Lewis structures of elements in the atomic structure chapter in Chemtutor. In this section there will be emphasis on the Lewis structure of small compounds and polyatomic ions.

This covalent bond between chlorine is one of the most covalent bonds known. Why? A covalent bond is the sharing of a pair of electrons. The two atoms on ether side of the bond are exactly the same, so the amount of "pull" of each atom on the electrons is the same, and the electrons are shared equally.

Next let's consider a molecule in which the atoms bonded are not the same, but the bonds are balanced. Methane, CH_4, is such a molecule. If there were just a carbon and a single hydrogen, the bond between them would not be perfectly covalent. In the CH_4 molecule, the four hydrogen atoms exactly balance each other out. The Lewis structure of methane does not have any electrons left over. The carbon began with four electrons and each hydrogen began with two electrons. Only the bars representing the shared pairs of electrons remain. The carbon now shares four pairs of electrons, so this satisfies the carbon's need for eight electrons in the outside shell. Each hydrogen has a single shared pair in the outside shell, but the outside shell of the hydrogen only has two electrons, so the hydrogen has a full outer shell also.

Carbons and hydrogens are nice and easy to write in Lewis structures, because each carbon must have four attachments to it and each hydrogen must have one and only one attachment to it. When the bonds around a carbon atom go to four different atoms, the shape of the bonds around that carbon is roughly tetrahedral, depending upon what the materials are around the carbon. Carbons are also able to have more than one bond between the same two. Consider the series ethane (C_2H_6), ethene (C_2H_4),(common name is ethylene), and ethyne (C_2H_2), (common name is acetylene).

H_3 –C –C –H_3 ethane H_2 –C = C–H_2 ethyne H–C = C–H acetylene

In writing the Lewis structure of compounds, the bars representing bonds are preferred to the dots representing individual electrons.

The double bars between the carbons in ethylene, C=C, represent a double bond between the two carbons, that is four shared electrons to make a stronger attachment between the two carbons. The triple bars between the carbons of acetylene represent a triple covalent bond between those two carbons, three pairs of shared electrons between those carbons. Every carbon has four bonds to it showing a pair of electrons to make eight electrons in the outer shell. Each hydrogen has one and only one bond to it for two electrons in the outer shell. All of the outer shells are filled.

While we are doing this, notice that the Lewis structure of a molecule will show the shape of the molecule. All of the bonds in ethane are roughly the tetrahedral angle, so all of the hydrogens are equivalent. This is true. The bonds in acetylene make it a linear molecule. The bonds in ethylene are somewhat trigonal around the carbons, and the carbons can not twist around that bond as they can around a single bond, so that the molecule has a flat shape and the hydrogens are not equivalent. This is also true. (You will see this in the study of organic chemistry. This type of difference between the positions of the hydrogens is called *cis - trans* isomerism.)

There is no issue of shape around the Group 1 elements. There is only one attachment to them, so no angle is possible around them. But there are some molecular compounds with only two atoms, such as nitrogen monoxide, NO. The only feature of this molecule is the bond between the nitrogen atom and the oxygen atom. The small difference in *electronegativity* between the oxygen and the nitrogen give the molecule a small *dipole*, a small separation of charge, so a small amount of *polarity*. (Because there are an odd number of electrons in NO, this makes for an interesting Lewis structure. Try it.) Iodine fluoride, IF, is another diatomic compound that should have some polarity. Diatomic molecules like chlorine gas, Cl_2, have no *electronegativity difference* (ΔEN) from side to side of the bond, so they are completely balanced and completely *non–polar*.

Group 2 elements have two electrons in the outer shell. Many of the compounds of Group 2 elements are ionic compounds, not really making an angle in a molecule. Molecules made with Group 2 elements that have two attached items to the Group 2 element have a linear shape, because the two attached materials will try to move as far from each other as possible. A *linear* shape means that a straight line could be made through all three atoms with the central element in the centre. The shape of carbon dioxide is linear with the carbon in the centre.

C = C = O

VSEPR stands for Valence Shell Electron Pair Repulsion. The idea is a disarmingly simple one. Electrons are all negatively charged, so they repel each other. If an atom has two electron groups around it, the electrons, and the atoms they are bounded to, are likely to be found as far as they can be from each other. "As far as they can get from each other," and still remain attached to the central atoms means that the angle around the central atom is 180°, a straight line. Molecules with two electron groups attached to a central atom have a linear electron group shape and a linear molecular shape. Unless there is a large difference in electronegativity from one side to the other of a linear compound, there is no separation of charge and no polar character of the molecular.

Covalent compounds with boron are good examples of *trigonal* shaped molecules. The *trigonal* shape is a flat molecule with 120° angles between the attached atoms. Again using the example of a boron atom in the centre, the attached elements move as far away from each other as they can, forming a *trigonal* shape, also called *triangular,* or *trigonal planner* to distinguish it from the *trigonal pyramidal* shape of compounds like ammonia. BF3, boron trifluoride, is an example of a molecule with a *trigonal planar shape*. Each fluorine atom is attached to the central boron atom. There are three bonds to the boron, so the *electron group shape* is *trigonal planar* around boron. The *molecular shape* is also trigonal planar in boron trifluoride because each electron group has a fluorine atom attached to it.

But, what if the central atom has two other atoms and a lone pair of electrons attached to it? Nitrogen oxychloride is an example of that. NOCl, is a molecule with nitrogen in the centre and an oxygen and a chlorine atom attached to the central nitrogen. When we go through the skeleton structure and distribute the electron dots, we find that there is a *double bond* between the nitrogen and the oxygen and a *lone pair* (unshared pair) of electrons on the nitrogen in addition to the *single bond* from the nitrogen to the chlorine. There are three *electron groups* around the nitrogen, making the *electron group shape* more or less *trigonal planar*. But only two of those electron groups have an atom attached, so the *molecular shape* of nitrogen oxychloride is *bent* or *angular*. NOCl is not a balanced shape, so it is likely that there is some separation of charge within the molecule, making it a somewhat *polar compound.*

Group 4 elements are not in the centre of a flat molecule when they have four equivalent attachments to them. As with two or three attachments, the attached items move as far as they can away from each other. In the case of a central atom with four things attached to it, the greatest angle between the attached items does not produce a flat molecule. If you were to cut off the vertical portion of a standard three-legged music stand so that it was the same length as the three legs, the angles among all four directions would be roughly equal. Try this with a gumdrop or a marshmallow. Stick four different colored toothpicks into the center at approximately the same angle. If you have done it right, the general shape of the device will be the same no matter which one of the toothpicks is up. This shape is called tetrahedral. The shape of a tetrahedron appears with the attached atoms at the points of the figure and each triangle among any three of them makes a flat plane. A tetrahedron is a type of regular pyramid with a triangular base.

Group 5 elements, for instance nitrogen or phosphorus, will become triple negative as they add three electrons in ionic reactions, but this is rare. Nitrides and phosphides do not survive in the presence of water. Covalent bonds with these elements do survive in water. From the Lewis structure of these elements in the previous section, you know that Group 5 elements have the capability of joining with three covalent bonds, but they don't make the trigonal shape because the Unshared pair of electrons acts like another bonded attachment. The shape of the bonds around nitrogen and phosphorus is tetrahedral, just like the bonds around Group 4 elements.

Group 6 elements, oxygen and sulfur, have two pairs of unshared electrons. Just as in Group 5 elements, these two pairs of unshared electrons serve as another attached atom for the shape of the molecule. Group 6 elements make tetrahedral molecules also, but now the items making the points of the tetrahedron are now limited to two. The angle between the hydrogens in water is about 105 degrees. This peculiar shape is one of the things that makes water so special.

Group 7 elements have only one chance of attachment, so there is no shape around these atoms.

Bonding Forces in Water

The alchemists of old had several other objectives aside from maing gold. The though of a fluid material that could dissolve anything, the universal solvent, was another alchemical project. No alchemist would say, though, what material would hold such a fluid. Surprisingly, the closest thing we have to a universal solvent is water. Water is not only a common material, but the range of materials it dissolves is enormous. The guiding principle for predicting which materials dissolve in which solvent is that 'like dissolves like.' Fluids in which the atoms are attached with covalent bonds will dissolve covalent molecules. Fluids with a separation of charge in the bonds will dissolve ionic materials.

The bods that hold hydrogen atoms to oxygen atoms are closer to covalent than ionic, but the bond does have a great deal of ioniccharacter. Oxygen atoms are more electronegative than hydrogn atoms, so the electron pair is held closer to the oxygen atom. Another way to look at it is that there are only a very sm number of water molecules ionized. The ionization of water, $H_2O \rightarrow H^+ + (OH)^-$, into hydrogen ions and hydroxide ions happeffect is quite important as the reason for the existence of acids and bases. Materials of a mildly covalent nature, such as small lcohols and sugars, are soluble in water due to the mostly covalent nature of the bonds in water.

The shpe of the water molecule is bent at about a 105 degree angle dueto the electron structure of oxygen. The two pairs of elecrons that force the attached hydrogens into something close to a etrahedral angle give the water molecule an unbalanced shape like a boomerang, with oxygen at the angle and the hydrogen atoms at the ends. We can think of the molecule has having an 'oxygen side' and a 'hydrogen side'. Since the oxygen atom pulls the electrons closer to it, the oxygen side of the molecule has a slight negative charge. Cations (positive ions) are attracted to the partial positive charge on the oxygen side of water molecules. Likewise, the hydrogen side of the molecule has a slight positive charge, attracting anions. Polar materials such as salts, materials that have a separation of charge, dissolve in water due to the charge separation of water. The origin of the separation is called a *dipole moment* and the molecule itself can be called a *dipole*.

Moleculesor atoms that have no center of asymmetry are non-polar. Atoms such as the inert gases have no center of asymmetry. Molecules such as methane, CH_4, are likewise totally symmetrical. Very small forces, called London forces, can be developed within such materials by the momentary asymmetries of the material and induction forces on neighboring materials. These small forces account for the ability of non-polar particles to become liquids and solids. The larger the atom or molecule, the more potent the London forces, possibly due to the greater ability to separate charge within a larger particle. The larger the inert gas, the higher its melting point and boiling point. In alkanes, a series of non-polar hydrocarbon molecules, the larger the molecule, the higher the melting and boiling point.

There may be London forces in water molecules, but the enormous force of the dipole interaction completely hides the small London forces. The dipole forces within water are particularly strong for two additional reasons. Dipole forces that involve hydrogen atoms around a strongly electronegative material such as nitrogen, oxygen, fluorine, or chlorine are particularly strong due to the small size of the hydrogen atom compared to the size of the dipole force. Such dipoles have significantly stronger forces, and have been called hydrogen bonds. In water, this effect is even greater due to the small size of the oxygen atom, thus the whole water molecule. In a water molecule hydrogen bonding is a large intermolecular force in a small volume on a small mass that makes it particularly noticeable.

Compare methane, CH_4, to water. They are similar in size and mass, but methane is non-polar and water is very highly polar due to the hydrogen bonding. The melting point for methane is -184 °C (89 K) and for water is 0 °C (273 K). The boiling point for methane is -161.5 °C (111.7 K) compared to water at 100 °C (373.2 K). The temperature range over which methane is a liquid is less than a quarter the range for water. Most of these differences are accountable from the hydrogen bonding of water. More about water later.

CHAPTER 9

Mols, Per cents, and Stoichiometry

Every chemist has dreamed that atoms were large enough to see and manipulate one at a time. The same chemist realizes after considering it, that if individual molecules were available for manipulation, it would take far too long to get anything done. The view from the atom is very different from the view of trillions and trillions of atoms. The mass action of the atoms that we see on our 'macro' view of the world is the result of the action of an incredibly large number of atoms averaged in their actions. The most usual way we count the atoms is by weighing them. The mass of material as weighed on a balance and the atomic weight of the material being weighed is the way we have of knowing how many atoms or molecules we are working with. Instead of counting eggs, we can count cartons of eggs, each carton of which has a given number, a dozen. Instead of counting B-B's, we can count liters of B-B's and find out how many B-B's are in a liter. Instead of counting oats, we buy train cars of oats and know the number of oats in a full train car.

There are less than one hundred naturally occurring elements. Each element has a characteristic atomic weight. Most

Periodic Charts include the atomic weight of an element in the box with the element. The atomic weight is usually not an integer because it is close to being the number of protons plus the average number of neutrons of an element. Let's use the atomic weight as a number of grams. This will give us the same number of any atom we choose. If we weigh out 1.008 grams of hydrogen and 35.45 grams of chlorine and 24.3 grams of magnesium, we will have the same number of atoms of each one of these elements. The neat trick with this system is that we can weigh the atoms on a grand scale of number of atoms and get a count of them. This number of atoms that is the atomic weight expressed in grams is Avogadro's number, 6.022 E 23. The name for Avogadro's number of Anything is a mole or mol. A mol of aluminum is 27.0 grams of aluminum atoms. Aluminum is an element, so the particles of aluminum are atoms. There are Avogadro's number of aluminum atoms in 27.0 grams of it. But 1.008 grams of hydrogen is Not a mol of hydrogen! Why not? Remember that hydrogen is one of the diatomic gases. There is really no such thing as loose hydrogen atoms. The total mass of a single hydrogen diatomic molecule (H2) is 2. 016 AMU. A mol of hydrogen gas has a mass of 2.016 grams. In that 2.016 gram mass is Avogadro's number of H2 molecules because that is the way hydrogen comes. A mol of water is 18.016 grams because each water molecule has two hydrogen atoms and one oxygen atom. A mol of water has in it Avogadro's number of water molecules. Another way to view the same thing is that a formula weight is the total mass of a formula in AMU expressed with units of grams per mol.

So Avogadro's number is just a number, like dozen or gross or million or billion, but it is a very large number. You could consider a mol of sand grains or a mol of stars. We are more likely to speak of a mol of some chemical, for which we can find the mass of a mol of the material by adding the atomic weights of all the atoms in a formula of the chemical. The unit of atomic weight or formula weight is grams/mol.

The chemical formula of a material should tell you:

(a) which elements are in the material;

(b) how many atoms of each element are in the formula;

(c) the total formula weight; and

(d) how the elements are attached to each other.

The symbols of the elements tell you which elements are in the material. The numbers to the right of each symbol tells how many atoms of that element are in the formula. The type of atoms and their arrangement in the formula will tell how the elements are attached to each other. A metal and a nonmetal or negative polyatomic ion shows an ionic compound. A pair of non-metals are bonded by covalent bonds. Some crystals have water of hydration loosely attached in the crystal. This is indicated by the dot such as in blue vitriol, $Cu(SO_4) \cdot 5H_2O$, showing five molecules of water of hydration to one formula of cupric sulfate.

The unit of the formula weight or molecular weight or atomic weight is "grams per mol," so it provides a relationship between mass in grams and mols of material.

$$nF_w = m$$

'n' is the number of mols, 'Fw is the formula weight, and 'm' is the mass.

PER CENTS BY WEIGHTS

All men weigh 200 pounds. All women weigh 125 pounds. What is the per cent by weight of woman in married couples? A married couple is one man and one woman. The total weight is 325 pounds. The formula for per cent is:

$$\frac{\text{Target}}{\text{Total}} \times 100\% = \text{Per cent}$$

In this case the woman is the target.

$$\frac{125\ \#}{325\ \#} \times 100\% = 38.461\% = 36.5\%$$

Notice that the units of pound cancel to make the per cent a pure number of comparison.

The weights of atoms are the atomic weights. What is the percentage of chloride in potassium chloride? The atomic weight of potassium is 39.10 g/mol. The atomic weight of chlorine is 35.45 g/mol. So the formula weight of potassium chloride is 74.55 g/mol. The chloride is the target and the potassium chloride is the total. 35.45 g/mol/x 100% = 47.55198 % or 47.6 % to three significant figures.

$$\frac{35.45\ g/mol}{74.55\ g/mol} \times 100\% = 47.5520\% = 47.6\%$$

You can do that with any part of a compound. What is the percentage of sulfate in beryllium sulfate tetrahydrate?

Notice that the examples here are done to two decimal points of the atomic weights. The problems in the practice bunch at the end of this chapter are done to one decimal point of the atomic weight.

BASIC STOICHIOMETRY

Pronounce stoichiometry as "stoy-kee-ah-met-tree," if you want to sound like you know what you are talking about, or "stoyk:," if you want to sound like a real geek. Stoichiometry is just a five dollar idea dressed up in a fifty dollar name. You can compare the amounts of any materials in the same chemical equation using the formula weights and the coefficients of the materials in the equation. Let's consider the equation for the Haber reaction, the combination of nitrogen gas and hydrogen gas to make ammonia.

$$N_2 + 3\ H_2 \rightarrow 2\ NH_3$$

The formula for nitrogen is N_2 and the formula for hydrogen is H_2. They are both diatomic gases. The formula for ammonia is NH_3. The balanced equation requires one nitrogen molecule and three hydrogen molecules to make two ammonia molecules, meaning that one nitrogen molecule reacts with three hydrogen molecules to make two ammonia molecules or one Mol of nitrogen and three Mols of hydrogen make two Mols of ammonia. Now we are getting somewhere. The real way we measure amounts is by weight (actually, mass), so 28 grams (14 g/mol times two atoms of nitrogen per molecule) of nitrogen and 6 grams of hydrogen (1 g/mol times two atoms of hydrogen per molecule times three mols) make 34 grams of ammonia. Notice that no mass is lost or gained, since the formula weight for ammonia is 17 (one nitrogen at 14 and three hydrogens at one g/mol) and there are two mols of ammonia made. Once you have the mass proportions, any mass-mass stoichiometry can be done by good old proportionation. What is the likelihood you will get just a simple mass-mass stoich problem on your test? You should live so long. Well, you should get one.

Rather than thinking in terms of proportions, think in mols and mol ratios, a much more general and therefore more useful type of thinking. A mol ratio is just the ratio of one material in a chemical equation to another material in the same equation. The mol ratio uses the coefficients of the materials as they appear in the balanced chemical equation. What is the mol ratio of hydrogen to ammonia in the Haber equation? 3 mols of hydrogen to 2 mols of ammonia. Easy. In the standard stoichiometry calculations you should know, All roads lead to Mols. You can change any amount of any measurement of any material in the same equation with any other material in any measurement in the same equation. That is powerful. The setup is similar to Dimensional Analysis.

1. Start with what you know (GIVEN), expressing it as a fraction.

2. Use definitions or other information to change what you know to mols of that material.

3. Use the mol ratio to exchange mols of the material given to the mols of material you want to find.
4. Change the mols of material you are finding to whatever other measurement you need.

How many grams of ammonia can you make with 25 grams of hydrogen? (Practice your mol math rather than doing this by proportion. Check it by proportion in problems that permit it.)

You are given the mass of 25 grams of hydrogen. Start there.

$$\left(\frac{25\,\mathrm{g\,H_2}}{1}\right)\left(\frac{\mathrm{mol}}{2.0\,\mathrm{g}}\right)\left(\frac{2\,\mathrm{mols\,NH_3}}{3\,\mathrm{mols\,H_2}}\right)\left(\frac{17\mathrm{g}}{\mathrm{mol}}\right) =$$

GIVEN Fw MOL RATIO Fw

25 g H_2/1 Change to mols of hydrogen by the formula weight of hydrogen 1 mol of H_2 = 2.0 g. (The 2.0 g goes in the denominator to cancel with the gram units in the material given.) Change mols of hydrogen to mols of ammonia by the mol ratio. 3 mols of hydrogen = 2 mols of ammonia. (The mols of hydrogen go in the denominator to cancel with the mols of hydrogen. You are now in units of mols of ammonia.) Convert the mols of ammonia to grams of ammonia by the formula weight of ammonia, 1 mol of ammonia = 17 g. (Now the mols go in the denominator to cancel with the mols of ammonia.) Cancel the units as you go.

The math on the calculator should be the last thing you do. 2 5 ÷ 2 . 0 × 2 ÷ 3 × 17 = and the number you get (141.66667) will be a number of grams of ammonia as the units in your calculations show. Round it to the number of significant digits your instructor requires (often three sig. figs.) and put into scientific notation if required. Most professors suggest that scientific notation be used if the answer is over one thousand or less than a thousandth. The answer is 142 grams of ammonia.

The calculator technique in the preceding paragraph illustrates a straightforward way to do the math. If you include all the numbers in order as they appear, you will have less chance of making an error. Many times students have been observed gathering all the numbers in the numerator, gathering all the numbers in the denominator, presenting a new fraction of the collected numbers, and then doing the division to find an answer. While this method is not wrong, the extra handling of the numbers has seen to produce many more errors.

Example below starts at "mass given" and goes throught the mol ratio to "mass find."

$mass_1 \rightarrow F_{w1} \rightarrow mol_1 \rightarrow$ mol ratio $2/1 \rightarrow mol_2 \rightarrow Fw_2 \rightarrow mass_2$

Notice by the chart above we may get the number of mols of material given if we change the mass by the formula weight, but in our continuous running math problem, we don't have to stop and calculate a number of mols. Students who insist on doing so tend to get more calculator errors.

The more traditional formula for converting mols to mass would be, where Fw is the formula weight, m is the mass, and n is the number of mols: n x Fw = m.

Density Times Mass of a Pure Material

Density multiplied by the volume of a pure material is equal to the mass of that material. If we know the density of a material and the volume of the pure material, with D = density and V = volume, DV = m so:

$D_1V_1 \rightarrow mass_1 \rightarrow F_{w1} \rightarrow mol_1 \rightarrow$ mol ratio
$2/1 \rightarrow mol_2 \rightarrow F_{w2} \rightarrow mass_2 \rightarrow 1/D_2 = V_2$

If you were given the density and volume of pure material you could calculate the volume of another material in that equation if you know it's density. Notice that the density must be inverted to cancel the units properly if you want the volume to find. If you need to find the density, the volume must be inverted.

Atoms or Molecules to Mols

One of the hardest ideas for some students is that the individual particles of a material are a single one of a formula of that material. Copper element comes only in the form of atoms. Water only comes in the form of a molecule with one oxygen and two hydrogen atoms. A mol, then is Avogadro's number of individual particles of whatever type of pure material the substance is made. There is no such thing as a mol of mud because mud is a mixture. There is no one mud molecule.

The word "pure" also can be misunderstood. We do not mean that a material is one hundred per cent the same material for us to use it, but that we are only considering the amount of that material.

The formula behind this relationship is: where n is the number of mols, A is Avogadro's number, and # is the number of individual particles of material,

$A \times n = \#$

Concentration Times Volume of a Solution

A *solution* is a mixture of a fluid (often water, but not always) and another material mixed in with it. The material mixed in with it is called the *solute*. There is more on solutions in the chapter devoted to that. The volume of a solution, V, is measured the same way the volume of a pure liquid is measured. The concentration can be expressed in a number of ways, the most common in chemistry is the M, molar. One molar is one mol of solute in a liter of fluid. It is important to notice that the fluid is usually nothing more than a diluting agent. For most of the reactions, the fluid does not participate in any reaction.

Concentration times volume is number of mols of the solute material.

$C \times V = n$

The "given" side of concentration times volume is easy. As with density times volume of a pure material, but the "find" side may need more work. You need one or the other of the concentration and volume before you can calculate the other. At the end of the Dimensional Analysis if you want concentration, you will be using the volume inverted. If you want the volume, you will be using the concentration inverted. This is not so difficult because the units will guide you.

GASES

Standard temperature is zero degrees Celsius. Standard pressure is one atmosphere. A mol of ANY gas at standard temperature and pressure (STP) occupies 22.4 liters. That number is good to three significant digits. The equation would be 1 mol gas = 22.4 L @STP. The conversion factor, the Molar Volume of Gas, is 1 mol gas/22.4 L @STP or 22.4 L @STP/1 mol gas.

Where n is the number of mols, V is the volume of a gas, and MVG is the molar volume of gas:

$$V = n \times MVG$$

Gases not at STP will require the Ideal Gas Law Formula,

$$P\,V = n\,R\,T$$

where P is the pressure of the gas in atmospheres, V is the volume of the gas in liters, n is the number of mols of gas, T is the Kelvin temperature of the gas, and R is the "universal gas constant" with the measurement of 0.0821 liter-atmospheres per mol-degree. We will have to do some algebra on the PV = nRT gas equation to do the gas portion of the stoichiometry problems.

In given we only need to solve for n. n = PV/RT. If we need to find the volume, pressure, or temperature of a gas, we need to solve for the unknown and include the "mols find" as the n. More about gases later.

The earmarks of a stoichiometry problem are: There is a reaction. (A new material is made.) You know the amount of one material and you are asked to calculate the amount of another matierial in the same equation.

CHAPTER 10

Reduction and Oxidation Reactions

Redox is the term used to label reactions in which the acceptance of an electron (reduction) by a material is matched with the donation of an electron (oxidation). A large number of the reactions already mentioned in the Reactions chapter are redox reactions.

Synthesis reactions are also redox reactions if there is an exchange of electrons to make an ionic bond. If chlorine gas is added to sodium metal to make sodium chloride, the sodium has donated an electron and the chlorine has accepted an electron to become a chloride ion or an attached chlorine.

If a compound divides into elements in a decomposition, a decomposition reaction could be a redox reaction. The electrolysis of water is a redox reaction. With a direct electric current through it, water can be separated into oxygen and hydrogen. H_2O width=20 $H_2 + O_2$ The oxygen and hydrogen in the water are attached by a covalent bond that breaks to make the element oxygen and the element hydrogen. Learning more about the conditions for redox reactions will show that the electrolysis of water is a redox reaction.

A single replacement reaction is always a redox reaction because it involves an element that becomes incorporated into a compound and an element in the compound being released as a free element.

A double replacement reaction usually is not a redox reaction.

Oxidation States

Before we go any further into redox, we must understand oxidation states. The idea of oxidation state began with whether or not a metal was attached to an oxygen. Unattached (free) atoms have an oxidation state of zero. Since oxygen almost always takes in two electrons when it is not a free element, the combined form of oxygen (oxide) has an oxidation state of minus two. The exception to a combined oxygen taking two electrons is the peroxide configuration. Peroxide can be represented by -O-O- where the each dash is a covalent bond and each 'O' is an oxygen atom. Peroxide can be written as a symbol, $(O_2)^{2-}$. The over-simplified way of showing this is that each oxygen atom has a negative one oxidation state, but that is not really so because the peroxides do not come in individual oxygen atoms. Peroxides are not as stable as oxides, and there are very many fewer peroxides in nature than oxides. H_2O_2 is hydrogen peroxide.

Hydrogen in compound always an oxidation state of plus one except as a hydride. A hydride is a compound of a metal and hydrogen. Hydrides react with water, so there are no hydrides found in nature. The formula XH or XH2 or XH3 or even XH4 where X is a metal is the general chemical formula for hydride.

The rules for oxidation state are in some ways arbitrary and unnatural, but here they are:

1. Any free (unattached) element with no charge has the oxidation state of zero. Diatomic gases such as O_2 and H_2 are also in this category.

2. All compounds have a net oxidation state of zero. The oxidation state of all of the atoms add up to zero.
3. Any ion has the oxidation state that is the charge of that ion. Polyatomic ions (radicals) have an oxidation state for the whole ion that is the charge on that ion. The ions of elements in Group I, II, and VII (halogens) and some other elements only have one likely oxidation state.
4. Oxygen in compound has an oxidation state of minus two, except for oxygen as peroxide, which is minus one.
5. Hydrogen in compound has an oxidation state of plus one, except for hydrogen as hydride, which is minus one.
6. In radicals or small covalent molecules, the element with the greatest electronegativity has its natural ion charge as its oxidation state.

A redox reaction will have at least one type of atom releasing electrons and another type of atom accepting electrons. How can you most easily tell if a reaction is redox? Label every atom on both the reactant and product side of the equation with its oxidation number. If there is a change in oxidation number from one side of the equation to the other of the same species of atom, it is a redox reaction. Each complete equation must have at least one atom species losing electrons and at least one atom species gaining electrons. The loss and gain of electrons will be reflected in the changes of oxidation number.

Let's take the following equation:

$$K_2(Cr_2O_7) + KOH \rightarrow 2\ K_2(CrO_4) + H_2O$$

Is it a redox equation or not? Potassium dichromate and potassium hydroxide make potassium chromate and water. Some of the atoms are easy. All of the oxygens in compound have an oxidation state of minus two. All of the hydrogens have an oxidation state of plus one. Potassium is a group one element, so it should have an oxidation state of plus one in the compounds. That seems to make sense because dichromate and chromate ions have a charge of minus two and there are two potassium atoms in each compound. Hydroxide ion has a charge of minus one and it has one potassium. But what about the chromium atoms? We can do a little primitive math on the material either from the starting point of the compound or the ion to find the oxidation state of chromium in that compound. The entire compound must have a net oxidation state of zero, so the oxidation numbers of two potassiums one chromium and four oxygens must equal to zero.

2 K + Cr + 4 O = 0

We know the oxidation state of everything else but the chromium.

2(+1) + Cr + 4 (-2) = 0 and Cr = +6. Or we could do it from the point of view of the chromate ion.

Cr + 4 O = -2

The oxygens are minus two each.

Cr + 4 (-2) = -2

Either way Cr = +6.

Now the dichromate;

2 K + 2 Cr + 7 O = 0 and 2 (+1) + 2 Cr + 7 (-2) = 0. Then 2 Cr = +12 and Cr = +6.

You can do the math for the dichromate ion to see for yourself that the chromium does not change from one side of this equation to the other. As suspicious-appearing as the equation might have seemed to you, it is not a redox reaction.

Consider copper metal in silver nitrate solution becomes silver metal and copper II nitrate. The oxygens do not change. Oxygen in compound is negative two on both sides. The nitrogen can not change. It does not move out of the nitrate ion where it has an oxidation state of plus five. (Is that right?) The other two have to change because they both are elements with a zero oxidation state on one side and in compound on the other. Silver goes from plus one to zero and Copper goes from zero to plus two.

$AgNO_3$ + Cu " s f" width=20 $Cu(NO_3)2$ + Ag (not a balanced reaction)

Think of this on a number line. The copper is oxidized because its oxidation number goes up from zero to plus two. The silver is reduced because its oxidation number reduces from plus one to zero.

Half Reactions

Consider the reaction: $AgNO_3 + Cu \rightarrow Cu(NO_3)_2 + Ag$ (not a balanced reaction)

Half reactions are either an oxidation or a reduction. Only the species of atom that is involved in a change is in a half reaction. In the above reaction, silver goes from plus one to zero oxidation state, but to account for everything, the electrons must be placed into the half reaction. $e^- + Ag^+$ Ag^0 (Reduction) Notice that the half reaction must be balanced in charge also and that the only way to balance it is to add electrons to the more positive side. The other half reaction is that of copper.

$Cu^0 \rightarrow Cu^{+2} + 2\ e^-$ (Oxidation)

This time the material is oxidized and the electrons must appear on the product side. We must double the silver half reaction to cancel out the electrons from right to left. The two half reactions can be added together to make one reaction, thus.

$2(e- + Ag^+ \rightarrow Ag^0)$

$Cu^0 \rightarrow Cu^{+2} + 2\ e-$

and the total reaction is:

$Cu^0 + 2Ag^+ \rightarrow Cu^{+2} + 2Ag^0$

In the complete reaction the number of electrons lost must equal the number of electrons gained. The number of electrons used in the reduction half reaction must equal the number of electrons produced in the oxidation half reaction. The entire half reactions must be multiplied by numbers that will equalize the numbers of electrons, and the final complete balanced chemical reaction must show these number relationships.

One of the important bits of information from adding the half reactions in this case is that the entire chemical equation will have to have two silver atoms for every copper atom in the reaction for the reaction to balance electrically. This type of information from the half reactions is sometimes the easiest or only way to balance a chemical equation. The redox balancing problems beginning with number 31 at the end of the chapter are good help for your further understanding.

From doing this math on a number of materials, you will find that it is possible to get some strange-looking oxidation states, to include some fractional ones. The oxidation state math works on fractional oxidation states also, even though fractional charges are not possible.

Reduction or Oxidation?

A reduction of a material is the gain of electrons. An oxidation of a material is the loss of electrons. This system comes from the observation that materials combine with oxygen in varying amounts. For instance, an iron bar oxidizes (combines with oxygen) to become rust. We say that the iron has oxidized. The iron has gone from an oxidation state of zero to (usually) either iron II or iron III. This may be difficult to remember. The easier way to tell if a half reaction is a reduction or oxidation is to plot the changing ion into the number line. If the oxidation state of the ion goes up the number line, it is an oxidation. If it goes down the number line, it is a reduction. Based on the KIS principle (Keep It Simple), remember only one rule for this.

CHAPTER 11

Gases

Introduction

Gases appear to us as material of very low density that must be enclosed to keep together. Unlike solids, gases have no definite shape. Unlike liquids, gases have no definite volume, but they completely fill a container. The volume of the container is the volume of the gas in it. A gas exerts a pressure on all sides of the container that holds it. Gas can be compressed by pressures greater than the pressure the gas on its container. The words vapor, fume, air, or miasma also describe a gas. Air describes the common mixture of gases in the atmosphere. A miasma is usually a bad-smelling or poisonous gas. The words vapor and fume suggest that the gas came from a particular liquid.

In the gaseous state matter is made of particles (atoms or molecules) that are not attached to each other. The intermolecular or interatomic forces that hold solids and liquids have been overcome by the motion of the molecules. The particles of a gas have too much thermal energy to stay attached to each other. The motion and vibration of the atoms pull the individual molecules apart from each other.

Liquid air (with all of the molecules touching each other) has a density of 0.875 grams per milliliter. By Avogadro's law, a mol of any gas occupies 22.4 liters at standard temperature and pressure (STP).

1 mol of any gas at STP = 22.4 litres

Air in the gas phase at standard temperature and pressure (1 atmosphere of pressure and 0°C.) has a mol of it (28.96 g) in 22.4 liters, coming to 1.29 grams per liter. Liquid air is over 675 times denser than the air at one atmosphere. As an estimate, each molecule of gas in the air has 675 times its own volume to rattle around in. Gases are mostly unoccupied space. Each molecule of a gas can travel for a long distance before it encounters another molecule. We can think of a gas as having a 'point source of mass', that is, the volume of the molecule is negligible compared to the space it occupies.

When a gas molecule hits another one, they bounce off each other, ideally in a completely elastic encounter. There is pressure within the gas that is caused by the gas molecules in motion striking each other and anything else in the gas. The pressure that a gas exerts on its container comes from the molecules of gas hitting the inside of the container and bouncing off.

There are some materials that do not appear in the form of a gas because the amount of molecular motion necessary to pull a molecule away from its neighbors is enough to pull the molecule apart. For this reason you are not likely to see large biological molecules such as proteins, fats, or DNA in the form of a gas.

The Ideal Gas Law Formula

A gas may be completely described by its makeup, pressure, temperature, and volume. Where P is the pressure, V is the volume, n is the number of mols of gas, T is the absolute temperature, and R is the Universal Gas Constant.

P V = n R T

This formula is the "Ideal Gas Law Formula." The formula is pretty accurate for all gases as we assume that the gas molecules are point masses and the collisions of the molecules are totally elastic. (A completely elastic collision means that the energy of the molecules before a collision equals the energy of the molecules after a collision, or, to put it another way, there is no attraction among the molecules.) The formula becomes less accurate as the gas becomes very compressed and as the temperature decreases. There are some correction factors for both of these factors for each gas to convert it to a Real Gas Law Formula, but the Ideal Gas Law is a good estimation of the way gases act. We will consider only the Ideal Gas Law Formula here. The Universal Gas Constant, R, can be expressed in several ways, depending upon the units of P, V, and T. One common R is 0.0821 litre - atmospheres per mol - degree. It is highly recommended that you know this value for R and the Ideal Gas Law Formula.

Variations on the Ideal Gas Law Formula

The Ideal Gas Law Formula is a wonderful place to begin learning almost all of the formulas for gases. You are not likely to get out of a chemistry class without a question like: What is the mass of neon in a neon light at 0.00545 Atmospheres at 24°s Celsius if the inside volume is 0.279 liters?

Given: P = 0.00545 Atmospheres T = 24°C + 273° = 297K V = 0.279 litres

Find: mass (m) of neon

m/Fw can now substitute for n and P V = (m/Fw) R T or Fw P V = m R T. When you solve for m, you almost have the problem completely done.

P V = (m/Fw) R T

The Combined Gas Low Formula

The Combined Gas Law Formula is the relationship of changing pressure, temperature, and volume of an ideal gas. The same amount of the same gas is given at two different sets of conditions. Let's call the first set of measurements, 'condition #1,' and the second set of measurements, 'condition #2.' We could label the pressure, temperature and volume symbols each with the subscripted number of the condition it represents. P1 is the pressure at condition #1. P2 is the pressure at condition #2. V_1 is the volume at condition #1, etc. The gas laws apply to both conditions, so $P_1 V_1 = n R T_1$ and $P_2 V_2 = n R T_2$. R is always the same Universal Gas Constant. If we are considering the same gas only at two different conditions, then $n_1 = n_2$. Since they are both equations, we could divide one equation by the other to get:

$$\frac{P_1V_1}{P_2V_2} = \frac{n_1RT_1}{n_2RT_2} \text{ or } \frac{P_1V_1}{P_2V_2} = \frac{T_1}{T_2} \text{ or } \frac{P_1V_1}{P_2V_2} = \frac{T_1}{T_2}$$

The last form can be a very useful one. This is the form of the Combined Gas Law Formula that Chemtutor finds easiest to remember. The formulas that most books call the Gas Laws are all contained in the Combined Gas Law. The Combined Law Formula is the one to use if you have any doubt about which of the Gas Laws to use.

$$\frac{P_1V_1}{P_2V_2} = \frac{T_1}{T_2}$$

Boyle's Law

Boyle's Law is useful when we compare two conditions of the same gas with no change in temperature. (Remember, "Always Boyle's at the same temperature!") No change in temperature means T1 = T2, so we can cancel the two temperatures in the Complete Gas Law Formula and get:

$$\frac{P_1 V_1}{P_2 V_2} = 1 \quad \text{or } P_1 V_1 = P_2 V_2$$

the usual Boyle's Law

$$P_1 V_1 = P_2 V_2$$

The usual expression of Boyle's Law was lurking right there in the Combined Gas Law Formula. As you can see, Boyle's Law is in the classic form of, "P is inversely proportional to V." We could predict that from the P and V being together in the numerator of the same side of the equation.

To get a feel for Boyle's Law, visualize a small balloon between your hands. The balloon is so small that you can push all sides of it together between your hands without any of the balloon pouching out at any point. When you push your hands together the volume of the gas in the balloon decreases as the pressure increases. When you let up on the pressure, the volume increases as the pressure decreases.

Charles's Law

Again we start with the Combined Law to get Charles's Law, but now there is no change in the pressure volume, so $P_1 = P_2$.

$$\frac{P_1 V_1}{P_2 V_2} = \frac{T_1}{T_2}$$

If you cancel out the two pressures, you get a form of Charles's Law that I consider easiest to remember. You can still see the P V = n R T in it if you look hard enough.

$$\frac{V_1}{V_2} = \frac{T_1}{T_2}$$

You may have seen this written differently, as in the following form:

$$\frac{V_1}{T_1} = \frac{V_2}{T_2}$$

These two expressions are mathematically exactly the same, but the first one shows its origin in the Combined Law. Remember it by, "Charles is under constant pressure."

To get a better feeling for Charles's Law, consider a child's toy balloon. At points between the beginning of filling of a balloon and the maximum stretching of a balloon, the change in internal pressure of a balloon is negligible as the balloon increases in size. A balloon is partially filled at room temperature and placed in the sun inside a car on a hot day in summer. The balloon expands in proportion to the Kelvin temperature. When the same balloon is take out of the car and put into a home freezer, the volume of the balloon decreases.

The Third Law

The third gas law from the Combined Gas Law has been named for Gay-Lussac in some books, Amonton in others, and not named in a large number of books. It is sometimes amusing to read a book that does not name the third law and needs to refer to it. The third law is the relationship of pressure and temperature with constant volume (V1 = V2.) the pressure and absolute temperature of a gas are directly proportional.

$$\frac{P_1 V_1}{P_2 V_2} = \frac{T_1}{T_2}$$

And so we get the third law, the relationship between the pressure and temperature of a gas.

$$\frac{P_1}{P_2} = \frac{T_1}{T_2}$$

Similarly to Charles's Law, it can be arranged so that it appears in the same form you see in most books.

$$\frac{P_1}{T_1} = \frac{P_2}{T_2}$$

To get a feel for the third Law, consider an automobile tire. With a tire gauge measure the pressure of the tire before and immediately after a long trip. When cool, the tire has a lower pressure. As the tire turns on the pavement, it alters its shape and becomes hot. There is some expansion of the air in the tire, as seen by the tire riding slightly higher, but we can ignore that small effect. If you were to plot the temperature versus pressure of a car tire, would zero pressure extrapolate out to absolute zero? Remember what you are measuring. The pressure of a car tire is actually the air pressure above atmospheric pressure. If you add atmospheric pressure to your tire gauge, you would certainly come closer to extrapolating to absolute zero.

Gas Stoichiometry Math

As you know from the Mols, per cents, and stoichiometry chapter section, stoichiometry is the calculation of an unknown material in a chemical reaction from the information given about another of the materials in the same chemical reaction. What if either the given material or the material you are asked to find is a gas? In stoichiometry you need to know the amount of one material. For gases not at STP, you must know the pressure, temperature, and volume to know the amount of material given. If you are given a gas not at STP, you will be able to substitute P V = n R T for the given side and plug it directly into the mols place by solving the equation for 'n'. Here is a sample problem using a gas not at STP as the given.

What mass of ammonia would you get from enough nitrogen with 689 liters of hydrogen gas at 350°C and 4587 mmHg?

Given: 689 l H2 = V T = 350°C + 273° = 623K P = 4587 mmHg (change to Atm)

Notice we have all three of the bits of data to know the amount of hydrogen.

Find: Mass (m) of NH_3

$$3\ H_2 + N_2 \rightarrow 2\ NH_3$$

(gas laws) → (mols given) → (mol ratio) → (formula weight find) → (mass find)

The ideal gas law (P v = n R T) must be solved for 'n' so it can be used as the 'given' of the outline.

$$\left(\frac{PV}{RT}\right)\left(\frac{\text{mols } NH_3}{\text{mols } H_2}\right)\left(\frac{\text{Fw } NH_3}{\text{mols } NH_3}\right) = \text{ammonia mass}$$

given mol ratio Fw find mass find

Things are a bit different when you need to find the volume, pressure, or temperature of a gas not at STP. You will need to solve P V = n R T for the dimension you need to find and attach it to the end of the sequence using the roadmap to find 'n' for the gas. Let's take another problem based on the same chemical equation to explore how to set up finding a gas not at STP.

What volume of ammonia at 7.8 atmospheres and 265°C would you get from 533 grams of nitrogen?

Given: m H2 = 533 g (Now hydrogen is the known material.)

Find: Volume of ammonia at P = 7.8 Atm and T = 265°C + 273° = 538K

The outline plan is now: (mass given) → (Fw given) → (mols given) → (mol ratio) → (gas laws)

$$\left(\frac{\text{mass of } H_2}{1}\right)\left(\frac{\text{mols } H_2}{\text{Fw of } H_2}\right)\left(\frac{\text{mols } NH_3}{\text{mols } H_2}\right) = \text{mols of ammonia}$$

given Fw given mol ratio mass find

Now the result of the stoichiometry is the number of mols of ammonia, 'n' in the ideal gas formula. We solve for the volume we want to find.

V = n R T/P and insert the numbers with 'n' coming from the stoichiometry, or we can tack (RT/P) onto the end of the stoichiometry.

$$\left(\frac{\text{mass of } H_2}{1}\right)\left(\frac{\text{mols } H_2}{\text{Fw of } H_2}\right)\left(\frac{\text{mols } NH_3}{\text{mols } H_2}\right)\left(\frac{RT}{P}\right) = V \text{ of } NH_3$$

given Fw given mol ratio gas low volume

Avogadro's Law

There is even more we can do with good old **P** V = n R T. The first part of this section introduced you to Avogadro's Law. One mole of any gas takes up a volume of 22.4 liters at standard temperature and pressure (STP). If we go back to the comparison of two formulas of the Ideal Gas Law, we have:

$$\frac{P_1V_1 = n_1RT_1}{P_2V_2 = n_2RT_2}$$

The R's are the same, so they can be cancelled. At standard temperature, $T_1 = T_2 = 273K$, and the T's can be cancelled. At standard pressure, P1 = P2 = 1 atmosphere, and the P's can be cancelled. When all the canceling has been done,

$$\frac{V_1}{V_2} = \frac{n_1}{n_2}$$

If the volume is proportional to the number of mols of a gas, there is a constant, k, that we can use in the formula, V = k n, to express the proportionality of V and n. What is that proportionality constant? At standard temperature and pressure, the pressure is one atmosphere and the temperature is 273K. The Universal Gas Constant is still 0.0821 Liter - atmospheres per mol - degree. Let's set n at one to find out what k is.

P V = n R T and V = n R T/P

V = (1 mol) (0.0821 L - A/ mol - K) (237 K)/(1 A)

Cancel the mols, the A's (for Atmosphere) and the K's. Do the math.

V = 22.4 Litres

We have seen this number before in Avogadro's Law, and this is where it comes from. When n is one mol and V is 22.4 Liters, k is 22.4 Litres/mol.

1 mol of any gas at STP = 22.4 liters

Dalton's Law of Partial Pressures

Similarly to the way we derived V = k n for Avogadro's Law above when the pressure is constant, we can derive P = k n for conditions when the volume does not change. This time there is no notable significance to the k, so we will just say that P is proportional to n when the temperature and pressure are constant. In conditions when more than one gas is mixed, we could number and add the pressures and mols. If we were to have P_1 of gas #1 due to n_1 mols of it and P_2 of another gas (#2) due to n_2 mols of it, those two gases in the same volume (They must be at the same temperature.) can be added together. PT is the total pressure and nT is the total number of mols.

$n_1 + n_2 = nT$ and $P_1 + P_2 = PT$

This has nothing to do with whether gas #1 is the same as gas #2. Dalton's Law of Partial Pressures says that, " The sum of all the partial pressures of the gases in a volume are equal to the total pressure." Where PT is the total pressure, P1 is the partial pressure of 'gas #1', P2 is the partial pressure of 'gas #2', Pn is the pressure of the last gas, whatever number (n) it.

Graham's Law of Diffusion (Or Effusion)

Gases under no change of pressure that either diffuse in all directions from an original concentration or effuse through a small hole move into mixture at a rate that is inversely proportional to the square root of the formula weight of the gas particle.

The mental picture of diffusion could be the drop of ink (with the same specific gravity as water) being carefully placed in the center of a glass of water. The ink will diffuse from the original point where it was deposited with no mixing of the glass of water. The mixing of diffusion is due to the movement of the molecules. Gases diffuse more quickly than liquids because the energy of motion is higher and the available path for unobstructed straight movement is much greater in gases.

Temperature is a type of energy. Temperature is the way we feel the motion of the molecules. $E = 1/2\ m\ v^2$ is the formula for energy of motion. This very motion of the molecules is the operating motion of the mixing action of diffusion. The mass of the molecule is the formula weight or molecular weight of the gas particle.

From the formula for energy of motion we can see that the mass of the particle (the formula weight) is inversely proportional to the square of the velocity of the particle. This is the easiest way to remember Graham's Law.

$$\frac{(v_1)^2}{(v_2)^2} = \frac{Fw_2}{Fw_1}$$

Notice in the above formula that 'v_1' is over 'v_2' and that 'Fw_2' is over 'Fw_1'. This is so that the inverse relationship can be expressed in the formula.

If you are solving for the effusion velocity of a particle, you might take the square root of both sides to get the other useful Graham's Law formula.

CHAPTER 12

Solutions

Introduction

A *solution* is a mixture of materials, one of which is usually a *fluid*. A fluid is a material that flows, such as a liquid or a gas. The fluid of a solution is usually the *solvent*. The material other than the solvent is the *solute*. We say that we *dissolve* the solute into the solvent.

Some solutions are so common to us that we give them a unique name. A solution of water and sugar is called *syrup*. A solution of sodium chloride (common table salt) in water is called *brine*. A sterilized specific concentration (0.15 molar) of sodium chloride in water is called *saline*. A solution of carbon dioxide in water is called *seltzer*, and a solution of ammonia gas in water is called *ammonia water*.

A solution is said to be *dilute* if there is less of the solute. The process of adding more solvent to a solution or removing some of the solute is called *diluting*. A solution is said to be *concentrated* if it has more solute. The process of adding more solute or removing some of the solvent is called *concentrating*. The *concentration* of a solution is some measurement of how much solute there is in the solution.

It might initially offend your sensibilities to consider a solution in which the solvent is a gas or a solid. The molecules of a gas do not have much interaction among them, and so do not participate to a large extent in the dissolving process. Solids are difficult to consider as solvents because there is a lack of motion of the particles of a solid relative to each other. There are, however, some good reasons to view some mixtures of these types as solutions. The molecules of a gas do knock against each other, and the motion of a gas can assist in vaporizing material from a liquid or solid state. The fan in a 'frost free' home freezer moves air around inside the freezer to sublimate any exposed ice directly into water vapor, a process clearly akin to dissolving. Solid metals can absorb hydrogen gas in a mixing process in which the metal clearly provides the structure.

Properties of Solutions

True solutions with liquid solvents have the following properties:

- The particles of solute are the size of individual small molecules or individual small ions. One nanometer is about the maximum diameter for a solute particle.
- The mixture does not separate on standing. In a gravity environment the solution will not come apart due to any difference in density of the materials in the solution.
- The mixture does not separate by common fiber filter. The entire solution will pass through the filter.
- Once it is completely mixed, the mixture is *homogeneous.* If you take a sample of the solution from any point in the solution, the proportions of the materials will be the same.
- The mixture appears clear rather than cloudy. It may have some color to it, but it seems to be transparent otherwise. The mixture shows no *Tyndall effect.* Light

is not scattered by the solution. If you shine a light into the solution, the pathway of the light through the solution is not revealed to an observer out of the pathway.

- The solute is completely dissolved into the solvent up to a point characteristic of the solvent, solute, and temperature. At a *saturation point* the solvent no longer can dissolve any more of the solute. If there is a saturation point, the point is distinct and characteristic of the type of materials and temperature of the solution.
- The solution of an ionic material into water will result in an *electrolyte* solution. The ions of solute will separate in water to permit the solution to carry an electric current.
- The solution shows an increase in osmotic pressure between it and a reference solution as the amount of solute is increased.
- The solution shows an increase in boiling point as the amount of solute is increased.
- The solution shows a decrease in melting point as the amount of solute is increased.
- A solution of a solid non-volatile solute in a liquid solvent shows a decrease in vapor pressure above the solution as the amount of solute is increased.

Last four of the properties of solutions collectively are called *colligative* properties. These characteristics are all dependent only on the number of particles of solute rather than the type of particle or the mass of material in solution.

Other Types of Mixture

Take a spoonful of dirt and vigorously mix it with a glass of water. As soon as you stop mixing, a portion of the dirt drops to the bottom. Any material that is suspended by the fluid motion alone is only in *temporary suspension*. A portion of the dirt makes a true solution in the water with all of the properties of the above table, but there are some particles, having a diameter roughly between 1 nm and 500 nm, that are suspended in a more lasting fashion. A suspended mixture of particles of this type is called a *colloid,* or *colloidal suspension,* or *colloidal dispersion.*

For colloids or temporary suspensions the phrase *dispersed material* or the word *dispersants* describes the material in suspension, analogous to the solute of a solution. The phrase *dispersing medium* is used for the material of similar function to a solvent in solutions.

As with true solutions, it is a bit of a stretch to consider solids as a dispersing medium or gases as forming a large enough particle to be a colloid, but most texts list some such. A *sol* is a liquid or solid with a solid dispersed through it, such as milk or gelatin. *Foams* are liquids or solids with a gas dispersed into them. *Emulsions* are liquids or solids with liquids dispersed through them, such as butter or gold-tinted glass. *Aerosols* are colloids with a gas as the dispersing medium and either a solid or liquid dispersant. Fine dust or smoke in the air are good examples of colloidal solid in a gas. Fog and mist are exampes of colloidal liquid in a gas.

Liquid dispersion media with solid or liquid dispersants are the most often considered. Homogenized whole milk is a good example of a liquid dispersed into a liquid. The cream does not break down into molecular sized materials to spread through the milk, but collects in small *micelles* of oily material and proteins with the more ionic or *hydrophilic* portions on the outside of the globule and the more fatty, or oily, or non-polar, or *hydrophobic* portions inside the ball-shaped little particle. Blood carries liquid lipids (fats) in small bundles called *lipoproteins* with specific proteins making a small package with the fat.

Proteins are in a size range to be considered in colloidal suspension in water. Broth or the independent proteins of blood or the casein (an unattached protein) in milk are colloidal. There are many proteins in the cellular fluids of living things that are in colloidal suspension.

Colloidal dispersants in water stay in suspension by having a layer of charge on the outside of the particle that is attractive to one end of water molecules. The common charge of the particles and the water *solvation layer* keep the particles dispersed. A *Cottrel precipitator* collects the smoke particles from air by a high voltage charge and collection device. Boiling an egg will denature and coagulate the protein in it. Proteins can be fractionally 'salted out' of blood by adding specific amounts of sodium chloride to make the proteins coagulate. The salt adds ions to the liquid that interfere with the dispersion of the colloidal particles.

Properties of Colloids

Colloids with liquid as a dispersing agent have the following properties:

- The particles of dispersant are the between about 500 nm to 1 nm in diameter.
- The mixture does not separate on standing in a standard gravity condition. (One 'g.')
- The mixture does not separate by common fiber filter, but might be filterable by materials with a smaller mesh.
- The mixture is not necessarily completely homogeneous, but usually close to being so.
- The mixture may appear cloudy or almost totally transparent, but if you shine a light beam through it, the pathway of the light is visible from any angle. This scattering of light is called the *Tyndall effect.*
- There usually is not a definite, sharp saturation point at which no more dispersant can be taken by the dispersing agent.

- The dispersant can be *coagulated*, or separated by clumping the dispersant particles with heat or an increase in the concentration of ionic particles in solution into the mixture.
- There is usually only small effect of any of the colligative properties due to the dispersant.

Concentration

The concentration of a solution is an indication of how much solute there is dissolved into the solvent. There are a number of ways to express concentration of a solution. By far the most used and the most useful of the units of concentration is *molarity*. You might see '6 M HCl' on a reagent bottle. The 'M' is the symbol for *molar*. One molar is one mol of solute per liter of solution. The reagent bottle has six mols of HCl per liter of acid solution. Since the unit 'molar' rarely appears in the math of chemistry other than as a concentration, to do the unit analysis correctly, you will have to insert concentrations into the math as 'mols per liter' and change answers of 'mols per liter' into molar.

Molality is concentration in mols of solute per kilogram of solvent. Mol fraction is the number of mols of solute per number of mols of solution. Weight-weight per cent (really mass per cent) is the number of grams of solute per grams of solution expressed in the form of a per cent. Mass-volume concentration is the number of grams of solute per milliliter of solution. There are other older units of concentration, such as Baumé, that are still in use, mainly in industrial chemicals.

Normality is the number of mols of effective material per liter. In acid-base titrations, the hydroxide ion of bases and the hydrogen (hydronium) ion of acids is the effective material. Sulfuric acid (H_2SO_4) has two ionizable hydrogens per formla of acid, or one mol of acid has two mols of ionizable hydrogen. 0.6 M H_2SO_4 is the same concentration as 1.2 N H_2SO_4.

We say that sulfuric acid is *diprotic* because it has two protons (hydrogen ions) per formula available. Hydrochloric acid (HCl) is *monoprotic*, phosphoric acid (H_3PO_4) is *triprotic*, and acids with two or more ionizable hydrogens are called polyprotic. Sodium hydroxide (NaOH) is *monobasic*, calcium hydroxide [$Ca(OH)_2$] is *dibasic*, and aluminum hydroxide is *tribasic*.

Where 'X' is the number of available hydrogen ions or hydroxide ions in an acid or base, N, the normality, is equal to the molarity, M, times X.

The normality system can be used for redox reactions, but the effective material is now available electrons or absorption sites for electrons. Consider the following reaction, #43 in the redox section.

In a sulfuric acid solution potassium permanganate will titrate with oxalic acid to produce manganese II sulfate, carbon dioxide, water, and potassium sulfate in solution.

+1 +7 -2 +1 +3 -2 +1 +6 -2 +2 +6–2 + 4 -2 +1 -2 +1 +6 - 2

$KMnO_4 + H_2C_2O_2 + H_2SO_4 \rightarrow MnSO_4 + CO_2 + H_2O + K_2SO_4$

5e– + Mn^{+7} * Mn^{+2} Reduction 5(C^{+3} * C^{+4} + e–) Oxidation

Balanced 2 $KMnO_4$ + 5 $H_2C_2O_4$ + 3 H_2SO_4 $\rightarrow$ 2 $MnSO_4$ + 10 CO_2 + 10 H_2O + 2 KCl

Since the manganese has a place for five electrons and the potassium permanganate contains the manganese, we could say that the normality of the permanganate solution is five times the molarity. The oxalic acid solution contains the carbon that becomes oxidized with only one electron added. The normality of the oxalic acid solution is the same as the molarity.

Where 'X' is the number of electrons either donated or accepted by a material in a redox reaction, the normality, N, is the molarity, M, times X.

In acid-base or redox titrations, the math is made easier by the use of normality. There is no need for a chemical equation because only the net reaction is considered. Where 'C' is the concentration in normality, and 'V' is the volume of solution, the formula is:

$$C_1 V_1 = C_2 V_2$$

Dissolving Solids into Liquids

The best way to measure the amount of a solid material is usually to weigh it. The best way to find the amount of a liquid is to find the volume. The formula for solutions is: C V = n, where C is concentration in molar, V is the volume in liters, and n is the number of mols of solute. Further, n = m/Fw , where m is the mass and Fw is the formula weight of the solute. Solving for the mass, m = C V Fw.

$$C V = n$$

or

$$m = C V Fw$$

How do you make the solution of a solid in a liquid. First weigh the solid to get the mass. The c vice such as a volumetric flask or a graduated cylinder. Use a small amount of water to dissolve the solute in the volumetric device. Add water to the volume desired and mix.

Expose the surface area of the solid to more solid an ve the solid. You can try this with sugar. Take two glasses of water at the same temperature and add a spoonful of sugar to each. Mix one, but not the other. In which glass does the sugar dissolve more easily?

Most solid materials will dissolve faster with increased temperature. Since the increased temperature increases the motion of the molecules, you can think of this effect as being similar to mixing. You have seen this effect. Sugar dissolves more quickly in warm tea than iced tea. Table salt dissolves more quickly in hot water than in cold.

How to Dissolve a Solid into a Liquid

Increase surface area of solid by decreasing the size of particle.

Increase temperature of the mixture.

Mix.

Dissolving Gases into Liquids

Gases are more easily measured by knowing the pressure, volume, and temperature of the gas. Seltzer water and ammonia water are two good examples of solutions of a gas in a liquid. Seltzer, or carbonated water, is the result of pressing carbon dioxide gas into water. Seltzer is used as the base liquid in any carbonated beverage. The bubbles in beer or sparkling wines are also due to carbon dioxide, but the CO_2 is a natural product of the fermentation process, so it does not have to be added artificially. Ammonia water, also called ammonium hydroxide solution, is made from ammonia (NH_3) being pressed into water. It is used as a weak base and as a cleaning material, particularly for glass.

Because the process is better done under pressure, it is often difficult to directly observe the actual dissolving done in most cases. The notable exception is the addition of dry ice, solid carbon dioxide, to water as described in the section on carbon dioxide.

As with a solid dissolving in a liquid, a gas dissolves in a liquid more easily with agitation or mixing, but that is where the similarity ends. Remove a carbonated beverage from its container and it becomes obvious that pressure is necessary to keep the gas in the liquid. The drink fizzes and bubbles, releasing the gas. As the beverage sits for a few hours, the taste becomes what we describe as 'flat.' Almost all of the carbon dioxide has escaped from the liquid. The only CO_2 remaining in the water will produce a partial pressure equal to the partial pressure of the gas in the atmosphere. Water carries dissolved oxygen from the partial pressure of the oxygen in the atmosphere.

As the combination of liquid and gas is NOT the favored (lowest energy) condition, an increase in temperature causes the separation. Lower temperature favors dissolving the gas into the liquid. You can verify this experimentally on your own. Leave one can of carbonated beverage at room temperature. Refrigerate a can of the same carbonated beverage. Gently heat a third can of the same beverage. Open them all and record the results. You are likely to find that the gas stays in solution better in the cooler liquid.

How to Dissolve a Gas into a Liquid

Increase the gas pressure on the liquid.

Decrease the temperature.

Mix.

Liquids in Liquids

A solution of two liquids is relatively uncomplicated. For the most part, liquids either mix together or they don't. When liquids will mix together, they do so in all proportions and are said to be *miscible*. If they do not mix, as oil and water, they are said to be *immiscible*. Using ethyl alcohol and water as examples of miscible liquids, we can have a solution of the two liquids with one drop of alcohol in a bucketful of water or one drop of water in a bucketful of alcohol.

Immiscible liquids can make a mixture of the nature of a colloidal suspension by very finely dividing one of the liquids and dispersing it through the other liquid. Milk fresh from the cow separates into a milk and a cream layer, the cream rising to the top. The cream of milk is a fatty material of a lower density, so it floats. The milk may be *homogenized*, a process that violently shakes the milk so that the cream forms very small ball-shaped particles. This homogenized milk will remain well mixed with normal treatment.

The stability of homogenized milk as a mixture is helped by the presence of the proteins of milk. Proteins often have areas of large amounts of available electrical charge and areas of very little charge. The areas of higher charge are more soluble in water and the areas of lower charge are more soluble in the fat of the cream. In this way the protein acts as a *surface active agent*, or *surfactant*. A surfactant is a large molecule with one area in one liquid and another area in another. Proteins of milk on the surface of the small globules of fat in homogenized milk will keep the globules from attaching back to each other, so the milk stays homogenized. Soaps and detergents are surfactants that help get oily dirt into suspension in water.

Agitation (mixing) is usually the most important factor in making a liquid-liquid mixture. The agitation of milk to homogenize it is a good example for colloids, but many other liquids do not mix without considerable agitation. If you make a highly concentrated syrup and pour it into water, the syrup will drop to the bottom of the water and stay there until it is agitated or (in a much longer time) diffusion mixes the layers.

Solubility

For the best view of solubility, we will use the examples of a solid solute dissolved into a liquid solvent. This does not mean that other materials do not work in the same fashion.

The solubility of a solution is a measure of how much of the solute can be dissolved into the solvent. The solution reaches a point called the *saturation point* when no more solute will be accepted by the solvent. Any further addition of solute will result in solid solute mixed in with the saturated solution. Each solvent and solute pair has a characteristic solubility at a given temperature. Usually as you increase the temperature, an increased amount of solute will be able to dissolve.

Take a Pyrex measuring cup and put in exactly a cup of table sugar. Heat water to boiling and pour in a small amount. Notice what happens. The volume of material in the cup appears to shrink! Continue adding boiling water until the level is back up to the 'one cup' mark. Notice the temperature of the solution. It takes heat to dissolve sugar. Stir. You should be able to almost dissolve all the sugar. The solution should be very close to the saturation point at that temperature. The solution should end up at about room temperature. Now add a few heaping tablespoons of sugar. Stir and attempt to dissolve all the sugar. If you succeed, add another few tablespoonsful of sugar. Put the saturated solution with a lot of undissolved sugar into the microwave, and heat until all the sugar is dissolved. If you have a meat thermometer, find the temperature of the boiling mixture. (Be careful. The solution is very hot. Handle with something to insulate you from the heat.)

Observe the solution after you take it out of the microwave and put it on the counter. Notice the temperature at which the sugar crystals begin to form again.

If you have done the experiment just right, you may see the crystals appearing at a temperature far below what you might think. If you boil the solution enough in the microwave, you will dissolve all traces of a seed crystal for the saturated solution to deposit sugar onto. At one time your solution will be *supersaturated*, or beyond the normal amount of solute in the solution. Supersaturation is an unstable condition. If any crystal is presented to a supersaturated solution, the crystallization of the solute onto it will occur fairly rapidly.

At home if you have done this demonstration with only sugar and water in a clean cup, don't waste the sugar solution. A little bit of maple flavoring will make it a fine syrup for pancakes, or you can use it in the frosting of the chocolate cake I have published here on the site. Do not eat any material made at school. Lab materials may contain traces of contaminants. If you eat anything in the school laboratory, the school lawyers will turn green and purple, have a conniption fit, and likely take their discomfort out upon you.

Solubility of salts depends upon the type of ions in the salt. There is a very great range of solubility of salts in water. Even the most insoluble, such as silver chloride, have a very small but detectable solubility. Some salts, called *deliquescent salts,* are so soluble that they grab water molecules out of the air and can dissolve themselves in this way.

Using the simplification of classifying materials as either soluble or not in water at room temperature, there are some nice easy general rules for predicting whether or not a salt will dissolve in water. These rules are useful not just for predicting how to make solutions, but ion reactions, such as a double displacement reaction, depend upon the insolubility of a salt as a possible product for the reaction to happen. Depending upon what your instructor suggests, it may be a good idea for you to know the following rules:

(a) Almost all simple ionic compounds with Group I elements or ammonium ion, $(NH_4)+$, are soluble.

(b) All nitrates (NO_4)-, most sulfates, $(SO_4)^{2-}$, and most chlorides, Cl-, are soluble. ** Notable exceptions to this rule are: barium sulfate, $BaSO_4)^{2-}$, lead II sulfate, $PbSO_4)^{2-}$, and silver chloride, AgCl.

(c) Most hydroxides, $(OH)^-$, carbonates, $(CO_3)^{2-}$ sulfides, S^{2-}, and phosphates, $(PO_4)^{3-}$, are insoluble except for the compounds of rule (a). Barium hydroxide, $Ba(OH)_2$ is a soluble exception to this rule.

Colligative Properties

The colligative properties of solutions have already been mentioned in the section on properties of solutions. A colligative property is one that depends only on the number of particles in solution rather than the type of particle. Molecular solutes have only one particle per formula, but ionic materials come apart into their ions and have almost as many particles in the solution as there are ions available. The word 'almost' was included on purpose because there is a small tendency for ions to re-associate with each other, making ion pairs that decrease the number of particles. The ion pair effect depends upon the properties of the species dissolved and the concentration of solute. The more concentrated the solute, the greater percentage of ion pairing takes place.

The colligative properties of solutions are:

1. The solution shows an increase in osmotic pressure between it and a reference solution as the amount of solute is increased.

   Osmotic pressure occurs when a semipermeable membrane divides two solutions, one of which has more solute than the other. A semipermeable membrane is one that lets through water but not any material in solution or in suspension in the water. Semipermeable membranes are an important part of any living thing. Cell membranes are semipermeable. The membranes on the outside of eggs are semipermeable. Trees pull up water from their roots by osmosis.

   Here is an easy way to demonstrate osmotic pressure. Take two similar hen's eggs and keep them in a dilute vinegar solution for a few days. The acid in the vinegar will react with the calcium compounds that are the hardening materials for the shell. There are two semipermeable membranes under the hard shell of the egg. Replace the vinegar solution if the process stops for a few days before all of the shell is removed. When all of the hard shell has gone, compare the size of the eggs. They should be fairly close. Put one egg into pure water (or tap water). Put the other egg into a brine solution (table salt dissolved in water). Observe the eggs over a few days.

Water goes through the semipermeable membrane in a direction to make the particle concentration on either side equal. The egg in just water will absorb water and become very large. The egg in brine will shrivel from water going out of it. The tight skin on the large egg is a demonstration of the pressure provided by osmosis.

Don't eat the eggs. Open them up and see what's inside. Inspect them carefully, particularly the yolk and its size. The membranes of the egg are a pretty good barrier to bacteria, but the stretched membrane particularly may not be able to keep bacteria out. Smell the eggs after you have opened them. Is there an odor that would indicate bacterial contamination? Cook them to see if the proteins react the same way as other eggs, but do not eat them due to the possibility of hidden bacterial contamination.

Red blood corpuscles (in humans) are not much more than semipermeable bags containing oxygen-absorbing protein (hemoglobin) floating in the blood. If you were to pump pure water into a person, the osmotic pressure due to the difference in *osmolarity* would swell and burst the red corpuscles. If the blood plasma has too many dissolved particles, the red corpuscles would shrivel up or *crenate*. Saline is a solution that is designed to be the same osmolarity as the cellular and corpuscular contents.

2. A solution of a solid non-volatile solute in a liquid solvent shows a decrease in vapor pressure above the solution as the amount of solute is increased.

Honey has some moisture in it that is close to saturation in sugar. Take two small shallow dishes and put in an equal (small) amount of honey in one and water in the other. Leave them exposed to the air in the same place, and observe them over a few days. The sugar in the honey will reduce the vapor pressure of the solution.

3. The solution shows an increase in boiling point as the amount of solute is increased.

   The boiling point of a liquid is just the point at which the vapour pressure of the liquid equals the surrounding pressure. If the vapor pressure decreases, it will take a greater temperature to boil the liquid.

   Put a small amount of honey in the bottom of a glass and about the same level of water in the same kind of glass. Place them both in a microwave oven. Which one boils first? Try the same experiment with various amounts of salt in solution.

4. The solution shows a decrease in melting point as the amount of solute is increased.

   It may be that the dissolved materials block the water molecules from attaching on to the rest of the water crystal. Or possibly that the dissolved material holds on to the water molecules more tightly than the water in the crystals.

   Whatever the cause, you have seen this in action in the making of homemade barrel ice cream. The barrel on the outside of the ice cream container has ice and salt (sodium chloride) in it. The ice melts (grabs up the heat) at a temperature lower than the usual melting point of water. Just ice in the barrel would not work, because it does not get cold enough to freeze the ice cream inside that has dissolved materials in it itself.

Concentration Math in Stoichiometry

If you are given the concentration and volume of a solution, you know the amount of solute in that solution. ($C\ V = n$) The concentration times volume can serve as the 'given' and will go directly to the mol ratio on the stoichiometry roadmap.

Since C V = n, and the first thing found from stoichiometry is the number of mols of a material (n), if you need to find the volume of a known concentration of a solution, you must attach (1/C) to the end of the roadmap to get the volume. If you need to find the concentration of a known volume of a solution, you must attach (1/V) to the end of the roadmap DA.

CHAPTER 13

Acids and Bases

Introduction

By the 1884 definition of Svante Arrhenius (Sweden), an acid is a material that can release a proton or hydrogen ion (H^+). Hydrogen chloride in water solution ionizes and becomes hydrogen ions and chloride ions. If that is the case, a base, or alkali, is a material that can donate a hydroxide ion (OH^-). Sodium hydroxide in water solution becomes sodium ions and hydroxide ions. By the definition of both Thomas Lowry (England) or J.N. Brønsted (Denmark) working independently in 1923, an acid is a material that donates a proton and a base is a material that can accept a proton. Was Arrhenius erroneous? $| 8-) No. The Arrhenius definition serves well for a limited use. We are going to use the Arrhenius definitions most of the time. The Lowry-Brønsted definition is broader, including some ideas that might not initially seem to be acid and base types of interaction. Every ion dissociation that involves a hydrogen or hydroxide ion could be considered an acid-base reaction. Just as with the Arrhenius definition, all the familiar materials we call acids are also acids in the Lowry-Brønsted model. The G.N. Lewis (1923) idea of acids and bases is broader than the Lowry-

Brønsted model. The Lewis definitions are: Acids are elecron pair acceptors and bases are electron pair donors.

We can consider the same idea in the Lowry- Brønsted fashion. Each ionizable pair has a proton donor and a proton acceptor. Acids are paired with bases. One can accept a proton and the other can donate a proton. Each acid has a proton available (an ionizable hydrogen) and another part, called the *conjugate base*. When the acid ionizes, the hydrogen ion is the acid and the rest of the original acid is the conjugate base. Nitric acid, HNO_3, *dissociates* (splits) into a hydrogen ion and a nitrate ion. The hydrogen almost immediately joins to a water molecule to make a hydronium ion. The nitrate ion is the conjugate base of the hydrogen ion. In the second part of the reaction, water is a base (because it can accept a proton) and the hydronium ion is its conjugate base.

HNO_3 +	H_2O →	NO_3^- +	H_3O^+
ACID	BASE	CONJUGATE BASE	CONJUGATE ACID

In a way, there is no such thing as a hydrogen ion or proton without anything else. They just don't exist naked like that in water solution. Remember that water is a very polar material. There is a strong partial negative charge on the side of the oxygen atom and a strong partial negative charge on the hydrogen side. Any loose hydrogen ion, having a positive charge, would quickly find itself near one of the oxygens of a water molecule. At close range from the charge attraction, the hydrogen ion would find a pair (its choice of two pairs) of unshared electrons around the oxygen that would be capable of filling the its outer shell. Each hydrogen ion unites with a water molecule to produce a *hydronium ion*, H^3O^+, the real species that acts as acid. The hydroxide ion in solution does not combine with a water molecule in any similar fashion. As we write reactions of acids and bases, it is usually most convenient to ignore the hydronium ion in favor of writing just a hydrogen ion.

Properties of Acids

For the properties of acids and bases we will use the Arrhenius definitions.

- Acids release a hydrogen ion into water (aqueous) solution.
- Acids neutralize bases in a neutralization reaction. An acid and a base combine to make a *salt* and water. A salt is any ionic compound that could be made with the anion of an acid and the cation of a base. The hydrogen ion of the acid and the hydroxide ion of the base unite to form water.
- Acids corrode active metals. Even gold, the least active metal, is attacked by an acid, a mixture of acids called 'aqua regia,' or 'royal liquid.' When an acid reacts with a metal, it produces a compound with the cation of the metal and the anion of the acid and hydrogen gas.
- Acids turn blue litmus to red. Litmus is one of a large number of organic compounds that change colors when a solution changes acidity at a particular point. Litmus is the oldest known pH indicator. It is red in acid and blue in base. The phrase, 'litmus test,' indicates that litmus has been around a long time in the English language. Litmus does not change color exactly at the neutral point between acid and base, but very close to it. Litmus is often impregnated onto paper to make 'litmus paper.'
- Acids taste sour. Tasting lab acids is not permitted by any school. The word 'sauer' in German means acid and is pronounced almost exactly the same way as 'sour' in English. (Sauerkraut is sour cabbage, cabbage preserved in its own fermented lactic acid. Stomach acid is hydrochloric acid. Although tasting stomach acid is not pleasant, it has the sour taste of acid. Acetic acid is the acid ingredient in vinegar. Citrus fruits such as lemons, grapefruit, oranges, and limes have citric acid in the

juice. Sour milk, sour cream, yogurt, and cottage cheese have lactic acid from the fermentation of the sugar lactose.

Properties of Bases

- *Bases release a hydroxide ion into water solution.* (Or, in the Lowry- Brønsted model, cause a hydroxide ion to be released into water solution by accepting a hydrogen ion in water.)
- *Bases neutralize acids in a neutralization reaction.* The word reaction is:

  Acid plus base makes water plus a salt. Symbolically, where 'Y' is the anion of acid 'HY,' and 'X' is the cation of base 'XOH,' and 'XY' is the salt in the product, the reaction is: HY + XOH $\rightarrow$ src = HOH + XY.
- *Bases denature protein.* This accounts for the "slippery" feeling on hands when exposed to base. Strong bases that dissolve in water well, such as sodium or potassium lye are very dangerous because a great amount of the structural material of human beings is made of protein. Serious damage to flesh can be avoided by careful used of strong bases.
- *Bases turn red litmus to blue.* This is not to say that litmus is the only acid- base indicator, but that it is likely the oldest one.
- *Bases taste bitter.* There are very few food materials that are alkaline, but those that are taste bitter. It is even more important that care be taken in tasting bases. Again, no school permits tasting of lab chemicals. Tasting of bases is more dangerous than tasting acids due to the property of stronger bases to denature protein.

Strong Acids and Strong Bases

The common acids that are almost one hundred per cent ionized are:

HNO_3 - nitric acid

HCl - hydrochloric acid

H_2SO_4 - sulfuric acid

$HClO_4$ - perchloric acid

HBr - hydrobromic acid

HI - hydroiodic acid

The acids on this short list are called *strong acids,* because the amount of acid quality of a solution depends upon the concentration of ionized hydrogens. You are not likely to see much HBr or HI in the lab because they are expensive. You are not likely to see perchloric acid because it can explode if not treated carefully. Other acids are incompletely ionized, existing mostly as the unionized form. Incompletely ionized acids are called *weak acids,* because there is a smaller concentration of ionized hydrogens available in the solution. Do not confuse this terminology with the concentration of acids. The differences in concentration of the entire acid will be termed *dilute* or *concentrated.* Muriatic acid is the name given to an industrial grade of hydrochloric acid that is often used in the finishing of concrete.

In the list of strong acids, sulfuric acid is the only one that is *diprotic,* because it has two ionizable hydrogens per formula (or two mols of ionizable hydrogen per mol of acid). (Sulfuric acid ionizes in two steps. The first time a hydrogen ion splits off of the sulfuric acid, it acts like a strong acid. The second time a hydrogen splits away from the sulfate ion, it acts like a weak acid.) The other acids in the list are *monoprotic,* having only one ionizable proton per formula. Phosphoric acid, H_3PO_4, is a weak acid. Phosphoric acid has three hydrogen ions available to ionize and lose as a proton, and so phosphoric acid is *triprotic.* We call any acid with two or more ionizable hydrogens *polyprotic.*

Likewise, there is a short list of strong bases, ones that completely ionize into hydroxide ions and a conjugate acid.

All of the bases of Group I and Group II metals except for beryllium are strong bases. Lithium, rubidium and cesium hydroxides are not often used in the lab because they are expensive. The bases of Group II metals, magnesium, calcium, barium, and strontium are strong, but all of these bases have somewhat limited solubility. Magnesium hydroxide has a particularly small solubility. Potassium and sodium hydroxides both have the common name of *lye*. Soda lye (NaOH) and potash lye (KOH) are common names to distinguish the two compounds.

LiOH - lithium hydroxide

NaOH - sodium hydroxide

KOH - potassium hydroxide

RbOH - rubidium hydroxide

CsOH - cesium hydroxide

$Mg(OH)_2$ - magnesium hydroxide

$Ca(OH)_2$ - calcium hydroxide

$Sr(OH)_2$ - strontium hydroxide

$Ba(OH)_2$ - barium hydroxide

The bases of Group I metals are all *monobasic*. The bases of Group II metals are all *dibasic*. Aluminum hydroxide is *tribasic*. Any material with two or more ionizable hydroxyl groups would be called *polybasic*. Most of the alkaline organic compounds (and some inorganic materials) have an amino group ($-NH_2$) rather than an ionizable hydroxyl group. The amino group attracts a proton (hydrogen ion) to become $(-NH_3)^+$. By the Lowry- Brønsted definition, an amino group definitely acts as a base, and the effect of removing hydrogen ions from water molecules is the same as adding hydroxide ions to the solution.

Memorize the strong acids and strong bases. Other acids or bases are weak.

Solubility and Dissociation

Now, after considering the bases of Group II metals, is a fine time to think about acidity of a solution and the solubility of the compound. Calcium and magnesium hydroxides are used in *antacids,* materials used to combat gastrointestinal acidity. How can that be if they are strong bases? In order to act as a base, the material must be dissolved. Almost all of these bases that are dissolved are dissociated, or ionized, but the low solubility of these bases makes them safe to swallow. An acid or base must first dissolve before it can dissociate (come apart) or ionize (become a pair of ions).

It is important to notice that just because a compound has a hydrogen or an OH group as a part of the structure does not mean that it can be an acid or a base. The hydrogens of methane, CH_4, are all very covalently attached to the carbon atom. Glycerin (or glycerol) has three OH groups in its structure.

CH_2OH
|
$CHOH$
|
CH_2OH

These are alcoholic OH groups attached to a carbon atom. The availability of the hydroxide or hydrogen as an Ion depends upon what it is attached to.

The chemical equation for the dissociation of nitric acid is:

$$HNO_3 \rightarrow (NO_3)^- + H^+$$

For strong acids and strong bases the equation goes completely to the right. There is none of the original acid or base, but only the ions of the material unattached to each other in the water.

Overview of pH

pH is just a handy way to express how acidic or alkaline a water solution is. The pH of a solution is the negative log of the hydrogen ion concentration. The hydrogen ion concentration is inversely proportional to the hydroxide ion concentration, and the two of them multiplied together give the number 1 E-14. The table below shows the relationship among these measurements at the integers.

Comments	$[H^+]$	pH	pOH	$[OH^-]$
Very base	E-14	14	0	E-0
	E-13	13	1	E-1
Base	E-12	12	2	E-2
	E-11	11	3	E-3
Slightly base	E-10	10	4	E-4
	E-9	9	5	E-5
	E-8	8	6	E-6
Neutral	E-7	7	7	E-7
	E-6	6	8	E-8
	E-5	5	9	E-9
Slightly acid	E-4	4	10	E-10
	E-3	3	11	E-11
Acid	E-2	2	12	E-12
	E-1	1	13	E-13
Very acid	E-0	0	14	E-14

The pH Box

The pH box is a similar sort of self-torture device to the temperature box. All four of the measurements are different ways to express exactly the same condition. The Kw of water, the dissociation constant, is a natural number amazingly close to 1 E-14. That is, when you multiply the hydrogen ion concentration $[H^+]$ by the hydroxide ion concentration $[(OH)^-]$ in pure water at near room temperature, the number is 1 E-14. If you know the

$[(OH)^-]$, you know the $[H^+]$ and *vice-versa*. These two measurements are not the same scale, but they are two different measurements of the same thing. The pH is just the negative log of the $[H^+]$ and the pOH is just the negative log of the $[(OH)^-]$. The final leg of the box is the relationship between the pH and pOH, and that is the easiest one. pH + pOH = 14 because this is the exponential form of the Kw equation.

The hardest part of working the pH box is doing the "number crunching." The math is easier on a scientific calculator. Only a masochist would think about trying to do the computations by hand with a log table.

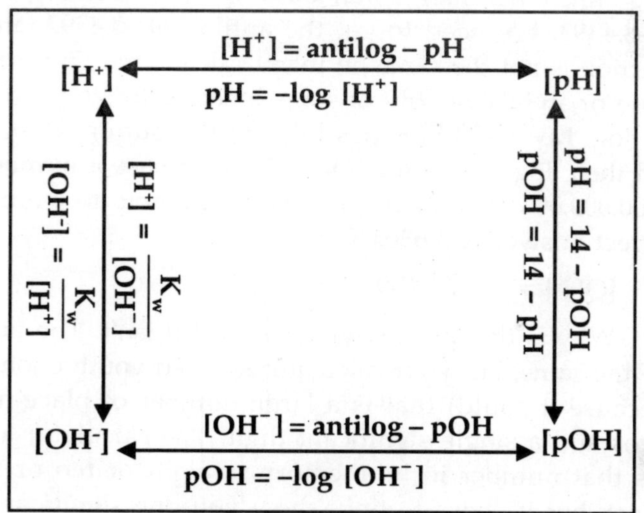

Fig. 13.1
The pH Box

Scientific Calculator use with pH Box

The calculator can be somewhat unfriendly in the math of the pH box. Let's take an example of using the box. The pH table before the pH box makes integer pH calculations easy, but the calculator is best used on non-integer pH's.

$[H^+]$ = 2.75 E-6

Start with $[H^+]$ = 2.75 E-6. Input 2 . 7 5 E +/- 6. To get the pH, punch 'log' (not *ln*, the natural log). The display shows -5.5607. The pH is the negative log, so it is 5.5607, rounded to 5.6, but you leave the -5.5607 on the display to keep going around the box.

pH = 5.5607

Punch + 1 4 = to add 14 to the negative pH. This will give you the pOH of 8.4393.

pOH = 8.4393

Punch the 'change sign' button, +/-. This changes 8.4393 to -8.4393. We need to get the antilog of -8.4393, and this function is not the same on many calculators. You may find a 2nd or an INV or shift or some other button to push before the log. My TI-30 SLR has INV on the button. Punch INV and then 'log' to get the $[OH^-]$. You may see a number like: 0.000,000,004 on your display. What did you do wrong? The correct answer is 3.6364 E-9.

$[OH^-]$ = 3.6364 E-9

Why is the display lying to you? It isn't. The numbers are the same, but your calculator showed you the long form (because it could) that is a large number of place-holding zeros and a single significant digit. The calculator actually has that number in its memory to eight or ten or sixteen digits, but it chose to only show you one significant digit. You can see the other digits on the display by multiplying by E6 (1 E6) or E9 (1 E9), but if you want to keep going around the box, you need to divide by the same number (with as many significant digits as you can) to get the $[OH^-]$ back. Or, you could store the $[OH^-]$ number before you take a look at it. Your calculator should have a button marked STO or M+ or M1 that will store your number into memory. Do that before you peek at the number. To get back that stored number, you punch RCL or M-. To get back to the original $[H^+]$, punch in: 1 E +/- 1 4 ÷ RCL = . You should see your good old $[H^+]$ of 2.75 E-6 on the display.

$[H^+]$ = 2.75 E-6

Now for practice, go around the pH box the other way.

The rules are:

- To get pH from $[H^+]$ or to get pOH from $[OH^-]$, use the negative of the log.
- To go from $[OH^-]$ to pOH or from $[H^+]$ to pH, use antilog of the negative number.
- To go from $[H^+]$ to $[OH^-]$ or back, first put in the Kw, 1E-14 and divide by the one you are leaving.
- To go from pH to pOH or back, subtract the number you have from 14.

Proficiency in pH box calculations requires practice. There is no Chemtutor Quickquiz on the pH box because you can make your own exercises and check your answers by coming back to the same number. You will use the pH box calculations in many problems in this acid- base section.

Weak Acids and Weak Bases

We can write the chemical equation for the dissociation of a weak acid, using 'A-' to represent the conjugate base, as:

$HA \rightarrow A^- + H^+$

And, similarly, we can write the chemical equation for the dissociation of a weak base, using 'X^+' to represent the conjugate acid, as:

$XOH \rightarrow OH^- + X^+$

The equilibrium expression for the dissociation of a weak acid is:

$$k_A = \frac{[H^+][A^-]}{[HA]}$$

In language, the equilibrium expression reads; "The dissociation constant of an acid is equal to the concentration of hydrogen ions times the concentration of the conjugate base of the acid divided by the concentration of un-ionized acid."

$$k_B = \frac{[X^+][OH^-]}{[XOH]}$$

Similarly, the equilibrium expression for a weak base reads: "The dissociation constant of a base equals concentration of hydroxide ions times concentration of conjugate acid divided by the concentration of un-ionized base."

The kA of an acid or the kB of a base are properties of that acid or base at the given temperature. The temperature at which these dissociation constants are listed is usually near room temperature.

The equilibrium expressions are for monoprotic acids or monobasic alkalis or the first dissociation of a polyprotic acid or a polybasic alkali. Phosphoric acid (H_3PO_4) is a good example of a polyprotic acid. When completely ionized, a mol of phosphoric acid will give three hydrogen ions and a phosphate ion, but the hydrogen ions come off one at a time at different pH's and with different kA's.

$H_3PO_4 \rightarrow (H_2PO_4)^- + H^+$ first ionization k_A = 6.92 E-3

$(H_2PO_4)^- \rightarrow (HPO_4)^{2-} + H^+$ second ionizaton k_A = 6.17 E-8

$(HPO_4)^{2-} \rightarrow (PO_4)^{3-} + H^+$ third ionization k_A = 2.09 E-12

Any acid with more than one ionizable hydrogen or any base with more than one ionizable hydroxide will usually separate stepwise as phosphoric acid.

The pK_A Box

The pK_A of an acid is a very useful number, as you will see in the math below. The pK_A is the negative log of the k_A, the pK_B is the negative log of the k_B, and the pK_A plus the pK_B equal fourteen. The k_A box is the same as the pH box, but substitute k_A for $[H^+]$, pK_A for pH, k_B for $[OH^-]$, and pK_B for pOH.

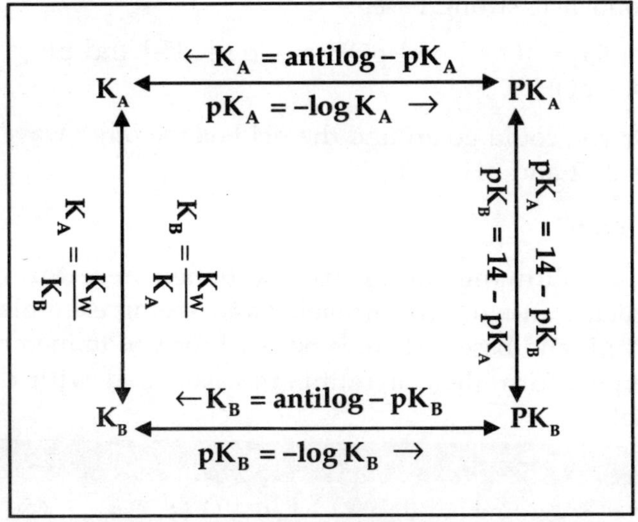

Fig. 13.2
pK_A Box

pH of Strong Acids and Bases

Strong acids and bases have all of the dissolved material completely ionized. The concentration of a monoprotic acid is equal to the concentration of hydrogen ion. The concentration of a monobasic alkali is equal to the concentration of hydroxide ion. The actual concentration of hydrogen ion (or hydroxide ion) from pure water is on the order of concentration of E-7 Molar, so any concentration of a strong acid or base over E-5 Molar completely swamps the comparatively tiny amount of ion from the ionization of water.

What is the pH of 0.0850 M HNO_3? Nitric acid is a monoprotic strong acid.

$[HNO_3] = [H^+]$ and pH = - log $[H^+]$, so, pH = - log (0.085) = 1.07

Only one step on the pH box.

What is the pH of 0.00765 KOH? Potassium hydroxide is a monobasic strong base.

[KOH] = [OH-] and pOH = - log [OH-] and pH = 14 - pOH

Or you could go around the pH box the other way. The pOH = 2.12 and pH = 11.88.

Weak Acids

The following tables are published here for your convenience in working problems and seeing examples of weak acids and bases. There is no need for you to memorize the names, formulas, or numbers associated with these materials.

Acid	Formula	k_A	pk_A
acetic acid	$H(C_2H_3O_2)$	1.74 E-5	4.76
ascorbic acid (1)	$H_2(C_6H_6O_6)$	7.94 E-5	4.10
ascorbic acid (2)	$(HC_6H_6O_6)^-$	1.62 E-12	11.79
boric acid (1)	H_3BO_3	5.37 E-10	9.27
boric acid (2)	$(H_2BO_3)^-$	1.8 E-13	12.7
boric acid (3)	(HBO_3)=	1.6 E-14	13.8
butanoic acid	$H(C_4H_7O_2)$	1.48 E-5	4.83
carbonic acid (1)	H_2CO_3	4.47 E-7	6.35
carbonic acid (2)	$(HCO_3)^-$	4.68 E-11	10.33
chromic acid (1)	H_2CrO_4	1.82 E-1	0.74
chromic acid (2)	$(HCrO_4)^-$	3.24 E-7	6.49
citric acid (1)	$H_3(C_6H_5O_7)$	7.24 E-4	3.14
citric acid (2)	$(H_2C_6H_5O_7)^-$	1.70 E-5	4.77

(Contd...)

Acid	Formula	k_A	pk_A
citric acid (3)	$(HC_6H_5O_7)=$	4.07 E-7	6.39
formic acid	$H(CHO_2)$	1.78 E-4	3.75
heptanoic acid	$H(C_7H_{13}O_2)$	1.29 E-5	4.89
hexanoic acid	$H(C_6H_{11}O_2)$	1.41 E-5	4.84
hydrocyanic acid	HCN	6.17 E-10	9.21
hydrofluoric acid	HF	6.31 E-4	3.20
lactic acid	$H(C_3H_5O_3)$	8.32 E-4	3.08
nitrous acid	HNO_2	5.62 E-4	3.25
octanoic acid	$H(C_8H_{15}O_2)$	1.29 E-4	4.89
oxalic acid (1)	$H_2(C_2O_4)$	5.89 E-2	1.23
oxalic acid (2)	$(HC_2O_4)^-$	6.46 E-5	4.19
pentanoic acid	$H(C_5H_9O_2)$	3.31 E-5	4.84
phosphoric acid (1)	H_3PO_4	6.92 E-3	2.16
phosphoric acid (2)	$(H_2PO_4)^-$	6.17 E-8	7.21
phosphoric acid (3)	$(HPO_4)=$	2.09 E-12	12.32
propanoic acid	$H(C_3H_5O_2)$	1.38 E-5	4.86
sulfuric acid (2)	$(HSO_4)^-$	1.05 E-2	1.98
sulfurous acid (1)	H_2SO_3	1.41 E-2	1.85
sulfurous acid (2)	$(HSO_3)^-$	6.31 E-8	7.20
uric acid	$H(C_5H_3N_4O_3)$	1.29 E-4	3.89

The organic acids in the table above have been written as acid with conjugate base as different from the standard notation. Notice that sulfuric acid's first ionization is a strong acid, so it is not on this table. Ascorbic acid is vitamin C. Some of the acids in the table are not very soluble, such as uric acid.

An organic acid usually has a - COOH group in it. The shape of that group is more like:

O = C OH
|

There is a double bond between the carbon and the single oxygen. There is a single bond with an alcohol group (-OH). There is one more bond to the central carbon that is usually attached to another carbon. The hydrogen of the alcohol group is the one that has a tendency to ionize away from the group, leaving a $(-COO)^-$ ion at the end of the acid.

Notice the set of organic acids that begin with formic acid. Formic acid has a hydrogen attached to the carbon in the organic acid group. Acetic acid has a $(-CH_3)$ group attached to the carbon in the organic acid group. The line of carbons becomes longer for the acids that follow. There are no branched chains of carbon atoms or any double or triple bonds between the carbons of the organic acids on this list.

The physiological organic acids (the organic acids that are found in living things) all have even numbers of carbons. Hexanoic, octanoic, and decanoic (C-6, C-8, and C-10 acids) are named for goats, which will give you an idea of the smell of those compounds.

Weak Bases

Many materials are weak bases due to the presence of an amino group $(-NH_2)$ attached to an organic compound. Nitrogen has five electrons in the outside shell. The three solo electrons can participate in (covalent) bonds to the nitrogen. The unshared pair of electrons can be donated by the nitrogen atom to constitute a complete covalent bond with a material that lacks a pair of electrons. The Lewis definition most clearly shows that the donor of the pair of electrons (the N atom) is a base and anything that attaches to the nitrogen by accepting a position of covalent attachment to the nitrogen is an acid.

Base	Formula	kB	pkB
alanine	$C_3H_5O_2NH_2$	7.41 E-5	4.13
ammonia (water)	NH_3 (NH_4OH)	1.78 E-5	4.75
dimethylamine	$(CH_3)2NH$	4.79 E-4	3.32
ethylamine	$C_2H_5NH_2$	5.01 E-4	3.30
glycine	$C_2H_3O_2NH_2$	6.03 E-5	4.22
hydrazine	N_2H_4	1.26 E-6	5.90
methylamine	CH_3NH_2	4.27 E-4	3.37
trimethylamine	$(CH_3)3N$	6.31 E-5	4.20

Alanine and glycine are amino acids, two of the twenty-or- so types of building blocks of protein. Each amino acid has both an amino (base) end and an acid end. Only the amino end (the base side) numbers are listed.

The 5% Rule

The measurement and calculation of pH is not as accurate as some other of the chemical measures due to differences in temperature, other ions present, purity of solutes, concentration changes due to evaporation, etc. Measurement of pH in medicine, for instance must be at 37 degrees Celsius. There may be machines that claim to measure pH to hundredths or thousandths of a pH unit, but the standard that calibrates the machine may be off a little. A change of five per cent in the hydrogen ion concentration will not change the pH more than a few hundredths of a pH unit. You try it on the calculator. Enter a number, say 0.001 and punch log. Now punch a number 95% or 105% of the first number (in this case, 0.00105 or 0.00095) and punch log. How different is the pH? Our answers in pH are going to be to the nearest tenth of a pH unit, so the hydrogen ion concentration needs to be only within 5% of an accurate number. This idea is the 'five per cent rule.'

You may remember we mentioned that the concentration of a strong acid is equal to the hydrogen ion concentration. That is not exactly so. In any water solution of an acid there is another source of hydrogen ions, the *water*. Water alone has a hydrogen ion concentration of E-7 Molar. Let's say you have a a E-5 M HCl solution, the contribution of the water is only a hundredth of the amount of hydrogen ion from the acid at the first approximation. You could approximate the real hydrogen ion concentration better by finding the hydroxide ion concentration of a solution with E-5 hydrogen ion concentration and then finding the added hydrogen ion concentration due to the dissociation of water from that. The total hydrogen ion concentration will be even further from having any significant contribution of hydrogen ion from the ionization of water. Does the ionization of water have an significance in this case? Of course not. There is much less than five per cent difference difference between the two numbers. Here is an obvious situation where you can use the simpler approximation of the hydrogen ion concentration to find the pH. There are some times when the simpler approximation is not accurate enough. There are some times when you may be in doubt and would need to work it BOTH ways to show whether you can use a simplified method.

pH of Weak Acids and Bases

Weak acids and bases do not ionize very much, so the $[H^+]$ or $[(OH)^-]$ must be calculated by the equilibrium expression.

$$K_A = \frac{[H^+][A^-]}{[HA]}$$

This would be easy to know exactly, except that the [HA] is not always equal to the concentration of solute that actually went into the solution. The more accurate way to express the equilibrium expression would be to include in the denominator the idea that the [HA] is really the concentration of solute originally put into solution minus the amount of that solute that ionized. We could represent the amount that ionized by either the $[A^-]$ or the $[H^+]$, so the denominator should be either $[HA] - [A^-]$ or $[HA] - [H^+]$. Let's use the second option for alternative denominator because we want to solve for the hydrogen ion concentration. By the 5% rule, $[HA] = [HA] - [H^+]$ only if the hydrogen ion concentration is less than five per cent of the total solute concentration. This is the case for most weak acids. The only exceptions are when the strongest of the weak acids are in the most dilute solutions.

If you have a stronger weak acid (one with a high k_A), you should check to see if the $[H^+]$ is near 5% of the total solute concentration. If it is, you will have to use the alternative denominator in the equation and solve it for $[H^+]$ by quadratic equation.

start with the ionization equilibrium expression with the alternative denominator

$$K_A = \frac{[H^+][A^-]}{[HA]-[H^+]}$$

simplify by combining $[A^-]$ and $[H^+]$ and then multiply by the denominator to get:

$$k_A[HA] - k_A[H^+] = [H^+]^2$$

and eventually, you get an expression that looks like a real quadratic equation in $[H^+]$ that can be solved by the quadratic formula

$$[H^+]^2 + k_A[H^+] - k_A[HA] = 0$$

Heaven forfend the need for using a quadratic equation. (I would rather kiss a BIG alligator.) It's a lot easier if the $[H^+]$ is less than five per cent of the [HA].

If there is only water and a weak acid in the solution, each mol of the acid dissociates into only one mol of hydrogen ions and only one mol of conjugate base ions. $[H^+] = [A^-]$! So;

$$k_A = \frac{[H^+]^2}{[HA]}$$

and solving for $[H^+]^2$,

$$[H^+]^2 = kA\,[HA]$$

then taking the square root of both sides, you have $[H^+]$.

$$[H^+] = \sqrt{kA\,[HA]}$$

Once you have the hydrogen ion concentration, you can go anywhere on the pH box.

The same idea goes for a weak base as the only material in water solution. Start with the equilibrium expression and the conjugate acid ion being equal to the hydroxide concentration. You will start with the base equilibrium expression:

$$K_B = \frac{[X^+][OH^-]}{[XOH]}$$

and solve for $[OH^-]$ to get;

$$[OH^-] = \sqrt{k_B\,[XOH]}$$

And it's back to the pH box for anything else you need.

Buffers and pH of Buffers

A buffer is a solution that resists changes in pH. A buffer is made with a weak acid and a soluble salt containing the conjugate base of the weak acid or a weak base and a soluble salt containing the conjugate acid of the weak base. Some examples of buffer material pairs are:

- acetic acid and sodium acetate;
- phosphoric acid and potassium phosphate;
- oxalic acid and lithium oxalate;
- carbonic acid and sodium carbonate; and
- or ammonium hydroxide and ammonium nitrate.

A buffer is most effective in solutions of pH at or close to the pKA of the weak acid. (Or solutions of pOH at or close to the pKB of the weak base.) The most buffering capacity is available when the concentration of weak acid or base is close to the concentration of the conjugate ion and when the concentration of both is greatest.

Thinking backwards, if you need a buffer at a particular pH lower than seven, chose the weak acid that has a pKA close to that pH and a conjugate base to go with it.

We can start again from the equilibrium expression of the ionization of a weak acid or a weak base;

$$k_A = \frac{[H^+][A^-]}{[HA]}$$

In buffer solutions the concentration of hydrogen ion is not equal to the concentration of conjugate ion. The concentration of conjugate ion from the dissociation of the acid is negligible, so we can calculate as if all the conjugate ion comes from the salt. If the soluble salt has only one mol of conjugate ion per mol of salt, the concentration of conjugate ion is the same as the concentration of salt.

The entrance to the pH box again is the hydrogen ion concentration. Solve for the hydrogen ion concentration if you need it, the pH of the solution, the pOH, or the hydroxide ion concentration.

$$[H^+] = \frac{K_A[HA]}{[A^-]}$$

which is the same as:

$$[H^+] = (k_A)\left(\frac{1}{[A]}\right)([HA])$$

The Henderson-Hasselbalch (H-H) equation can be somewhat confusing, but it is nothing more than the negative log of the the equation above, the equilibrium expression solved for the hydrogen ion concentration. If you know the equilibrium equation, know how to solve the equilibrium equation for the hydrogen ion concentration, and know how to convert the hydrogen ion concentration to the pH, you have it licked. If you really need the H-H equation for a test, it is best to make sure you are getting it right by deriving it from the equilibrium expression.

$-\log[H^+] = pH$

$-\log k_A = pKA$

$-\log(1/[A^-]) = +\log[A^-]$

$-\log([HA]) = -\log[HA]$

Component-by-component take the negative log of the equilibrium expression and change the multiplication to addition. (They are logs now.) You get:

$pH = pK_A + \log[A^-] - \log[HA]$,

which is the same as;

$pH = pK_A - \log([HA]/[A^-])$,

the usual way you see the H-H equation.

There are four perfectly correct ways to write the H-H equation. They are:

$$pH = pK_A - \log\left(\frac{HA}{[A^-]}\right)$$

$$pH = pK_A + \log\left(\frac{A^-}{[HA]}\right)$$

$$pOH = pK_B + \log\left(\frac{\text{conjugatecation}}{[\text{base}]}\right)$$

$$pOH = pkB - \log\left(\frac{\text{base}}{[\text{conjugatecation}]}\right)$$

An *equimolar* buffer is one in which the concentration of the weak acid (or base) is the same as the concentration of the conjugate ion. This may not seem particularly significant to you, but there are several important ideas that can be easily seen from it. Start with the ionization equilibrium expression and cancel the [HA] with the [A⁻]. This shows that in an equimolar buffer kA = [H⁺].

$$k_A = \frac{[H^+][A^-]}{[HA]}$$

if $[HA] = [A^-]$

$$k_A = [H^+]$$

This shows how a buffer works. As acid or base is added to a buffer solution, the buffer equilibrium will trade off [HA] with [A⁻] to come to a new equilibrium. When that happens, there will be a much smaller change in the hydrogen ion concentration than if the acid or base were added to an unbuffered solution.

Now let's do a similar trick with the H-H equation. If $[HA] = [A^-]$, the term $[HA]/[A^-] = 1$ and $\log 1 = 0$, so pH = pKA.

$$pH = pK_A - \log \frac{[HA]}{[A^-]}$$

$$pH = pK_A$$

This indicates that in an equimolar buffer the $pH = pK_A$ and that the ratio of the [HA] to the [A⁻] will determine the pH of the solution. The further the ratio gets from one to one, the further the pH gets from the pK_A. The buffer has its greatest buffering power at the pK_A of the weak acid (or base). The higher the concentration of both the weak acid and its conjugate ion, the more buffering power is available.

Titation

The word "titration" rhymes with "*tight nation,*" and refers to a commonly used method of (usually) finding the concentration of an unknown liquid by comparing it with a known liquid. An acid-base titration is good to consider when learning the method, but there are more uses for the technique. The measure of oxalate ion using potassium permanganate in a warm acid environment is a good example of a redox titration. The Mohr titration is a determination of chloride concentration using known silver nitrate solution and sodium dichromate an indicator.

A measured amount of the unknown material in a flask with indicator is usually combined with the known material from a buret (rhymes with "sure bet"). The buret is marked with the volume of liquid by a scale with zero on top and (usually) fifty milliliters on the bottom. The buret has some type of valve at the bottom that can dispense the contained liquid.

It is not necessary to start the titration with the known liquid level in the buret at the zero mark, but the level must be within the portion of the buret that is marked. The buret on the left shows about 1.8 ml of the yellow liquid in it. Most laboratory burets can be read to an accuracy of one hundredth of a milliliter. (The drawing on the left is a bit crude. Most burets show the ten divisions of a milliliter and you can interpolate between the marks.) One reads the buret by getting at eye level to the bottom of the meniscus (curve in the liquid) and comparing the bottom of the meniscus to the marks on the glass. A reading of the buret is taken before and at the end of the titration. The amount of know-concentration liquid used is the difference of the beginning and ending buret reading.

The *endpoint* of the titration is usually shown by some type of indicator. A *pH indicator* is a material, usually an orgainc dye, that is one color above a characteristic pH and another colour below that pH. There are many materials that can serve as pH indicators, each with its own ph range at which it changes color. Some have more than one color change at distinct pH's. Litmus and phenolphthalein are common pH indicators. Litmus is red in acid (below pH 4.7) and blue in base (above pH 8.1). Phenolphthalein (The second 'ph' is silent and the 'a' and both 'e's are long, if that is any help.) is clear in acid (below pH 8.4) and pink- purple in base (above pH 9.9). These ranges may seem large, but near the *equivalence point,* the point at which the materials are equal, there is a large change in pH. The equivalence point may not occur at pH 7, neutral pH, so the appropriate pH indicator must be chosen for the type of acid and base being titrated.

The volume of the material of unknown concentration is known by how much is put into the reaction vessel. The concentration of the standard is known, and its volume is known from the measurement of liquid used in the titration.

If you have a monobasic base and a monoprotic acid, you can simplify the formula for titration to: if C_A = the concentration of the acid and C_B = the concentration of the base and V_A = the volume of acid solution and V_B = the volume of base, then, $C_A V_A = C_B V_B$

Salts

A *salt* is the combination of an anion (- ion) and a cation (a + ion). Another way to think of a salt is the combination of the anion of a certain acid combined with the cation of a certain base. The neutralization of potassium hydroxide with hydrochloric acid produces water and the salt, potassium chloride. In a solid salt, the ions are held together by the difference in charge. Solid salts usually make crystals, sometimes including specific molar amounts of water, called *water of hydration* into the crystal. If a salt dissolves in water solution, it usually dissociates (comes apart) into the anions and cations that make up the salt.

Salts dissolved in water may not be at neutral pH. Table salt, NaCl, has a neutral pH in water, but baking soda, $NaHCO_3$ is very alkaline when dissolved in water. Take a teaspoonful of baking soda on your hand, wet it, and completely wash your hands with it. (Baking soda is a really fine material for cleaning hands!) There is the hint of the slipperiness of bases on your hands that had to come from the baking soda.

How can you predict the pH of a salt dissolved in water? The actual pH will depend on the type of anion and cation, the solubility of the salt, the temperature of the solution, and concentration of the salt if less than saturated. salt. Here are the general rules:

- Salts made of the anion of a strong acid and the cation of a strong base will be neutral salts, that is, the water solution with this salt will have a pH of seven. (example - sodium chloride)
- Salts made of the anion of a strong acid and the cation of a *weak* base will be acid salts, that is, the water solution with this salt will have a pH of less than seven. (example - ammonium chloride)
- Salts made of the anion of a *weak* acid and a strong base will be an alkali salt. The pH of the solution will be over seven. (example - sodium bicarbonate)
- It can be a bit more difficult to tell the pH of a salt solution if the salt is made of the anion of a weak acid and the cation of a weak base. Usually, the main determining factor is whether the weak acid is weaker than the weak base, but that is not always the case. For the purpose of the problems in the review section, you may say that the pH of a 'weak-weak' salt is indeterminate.

CHAPTER 14

Basic Concepts in Chemistry

Chemical nomenclature is the term given to the naming of compounds. Chemists use specific rules and "conventions" to name different compounds. This section is designed to help you review some of those rules and conventions.

Oxidation and Reduction

When forming compounds, it is important to know something about the way atoms will react with each other. One of the most important manners in which atoms and/or molecules react with each other is the oxidation/reduction reaction. Oxidation/Reduction reactions are the processes of losing and gaining electrons respectively. Just remember, "LEO the lion says GER:" Lose Electrons Oxidation, Gain Electrons Reduction. Oxidation numbers are assigned to atoms and compounds as a way to tell scientists where the electrons are in a reaction. It is often referred to as the "charge" on the atom or compound. The oxidation number is assigned according to a standard set of rules. They are as follows:

- An atom of a pure element has an oxidation number of zero.
- For single atoms in an ion, their oxidation number is equal to their charge.

- Fluorine is always -1 in compounds.
- Cl, Br, and I are always -1 in compounds except when they are combined with O or F.
- H is normally +1 and O is normally -2.
- The oxidation number of a compound is equal to the sum of the oxidation numbers for each atom in the compound.

Forming Ionic Compounds

Knowing the oxidation number of a compound is very important when discussing ionic compounds. Ionic compounds are combinations of positive and negative ions. They are generally formed when nonmetals and metals bond. To determine which substance is formed, we must use the charges of the ions involved. To make a neutral molecule, the positive charge of the cation (positively-charged ion) must equal the negative charge of the anion (negatively-charged ion). In order to create a neutral charged molecule, you must combine the atoms in certain proportions. Scientists use subscripts to identify how many of each atom makes up the molecule. For example, when combining magnesium and nitrogen we know that the magnesium ion has a "+2" charge and the nitrogen ion has a "-3" charge. To cancel these charges, we must have three magnesium atoms for every two nitrogen atoms:

$$3Mg^{2+} + 2N^{3-} \rightarrow Mg_3N_2$$

Knowledge of the charges of ions is crucial to knowing the formulas of the compounds formed.

- alkalis (1st column elements) form "+1" ions such as Na^+ and Li^+
- alkaline earth metals (2nd column elements) form "2+" ions such as Mg^{2+} and Ba^{2+}
- halogens (7th column elements) form "-1" ions such as Cl^- and I^-

Other common ions are listed in the table below:

Positive ions (cations)	Negative ions (anions)
1^{+}	1^{-}
ammonium ($NH4^{+}$)	acetate ($C_2H_3O_2^{-}$)
copper(I) (Cu^{+})	azide ($N3^{-}$)
hydrogen (H^{+})	chlorate (ClO_3^{-})
silver (Ag^{+})	cyanide (CN^{-})
	dihydrogen phosphate ($H2PO_4^{-}$)
2^{+}	hydride (H^{-})
cadmium (Cd^{2+})	bicarbonate (HCO_3^{-})
cabalt(II) (Co^{2+})	hydroxide (OH^{-})
copper(II) (Cu^{2+})	nitrate (NO_3^{-})
iron (Fe^{2+})	nitrite (NO_2^{-})
lead (Pb^{2+})	perchlorate (ClO_4^{-})
manganese(II) (Mn^{2+})	permanganate (MnO_4^{-})
mercury(I) (Hg_2^{2+})	thiocyanate(SCN^{-})
mercury(II) (Hg^{2+})	
nickel (Ni^{2+})	2^{-}
tin (Sn^{2+})	carbonate (CO_3^{2-})
zinc (Zn^{2+})	chromate (CrO_4^{2-})
	dichromate ($Cr_2O_7^{2-}$)
3+	hydrogen phosphate (HPO_4^{2-})
aluminum (Al^{3+})	oxide (O^{2-})
chromium(III) (Cr^{3+})	peroxide (O_2^{2-})
iron(III) (Fe^{3+})	sulfate (SO_4^{2-})
	sulfide (S^{2-})
	sulfite (SO_3^{2-})
	3^{-}
	nitride (N^{3-})
	phosphate (PO_4^{3-})
	phosphide (P^{3-})

Naming Ionic Compounds

The outline below provides the rules for naming ionic compounds:

Positive Ions

Monatomic cations (a single atom with a positive charge) take the name of the element plus the word "ion"

Examples:

- Na^{+} = sodium ion
- Zn+2 = zinc ion

If an element can form more than one (1) positive ion, the charge is indicated by the Roman numeral in parentheses followed by the word "ion"

Examples:

- Fe^{2+} = iron(II) ion
- Fe^{3+} = iron (III) ion

Negative Ions

Monatomic anions (a single atom with a negative charge) change their ending to "-ide"

Examples:

- O^{2-} = oxide ion
- Cl^{-} = chloride ion

Oxoanions (negatively charged polyatomic ions which contain O) end in "-ate". However, if there is more than one oxyanion for a specific element then the endings are:

- Polyatomic anions (a negatively charged ion containing more than one type of element) often add a hydrogen atom; in this case, the anion's name either adds "hydrogen-" or "bi-" to the beginning.

Two less oxygen than the most common starts with "hypo-" and ends with "-ite"	One less oxygen than the most common ends with "-ite"	THE MOST COMMON OXOANION ENDS WITH "-ATE"		One more oxygen than the most common starts with "per-" and ends with '-ate"
		Most common oxyanions with four oxygens	Most common oxyanions with three oxygens	
ClO^- = hypochlorite	• ClO^{2-} = chlorite • NO_2^- = nitrate • SO_3^{2-} = sulfate	• SO_4^{2-} = sulfate • PO_4^{3-} = Phosphate • CrO_4^{2-} = chromate	• NO_3^- = nitrate • ClO_3^- = chlorate • CO_3^{2-} = carbonate	ClO_4^- = perchlorate

Example:

CO_3^{2-} becomes HCO_3^-

"Carbonate" becomes either "Hydrogen Carbonate" or "Bicarbonate"

When combining cations and anions into an ionic compound, you always put the cation name first and then the anion name (the molecular formulas are also written in this order as well.)

Examples:

- $Na^+ + Cl^- \rightarrow NaCl$

  sodium + chloride → sodium chloride

- $Cu^{2+} + SO_4^{2-} \rightarrow CuSO_4$

  copper(II) + sulfate → copper(II) sulfate

- $Al^{3+} + 3NO_3^- \rightarrow Al(NO_3)_3$

  aluminum + nitrate → aluminum nitrate

Arrangement of Atoms

In naming ions, it is important to consider "isomers." Isomers are compounds with the same molecular formula, but different arrangements of atoms. Thus, it is important to include some signal within the name of the ion that identifies which arrangement you are talking about. There are three main types of classification, geometric, optical and structural isomers.

Geometric Isomers

It refers to which side of the ion atoms lie. The prefixes used to distinguish geometric isomers are *cis* meaning substituents lie on the same side of the ion and *trans* meaning they lie on opposite sides. Below is a diagram to help you remember.

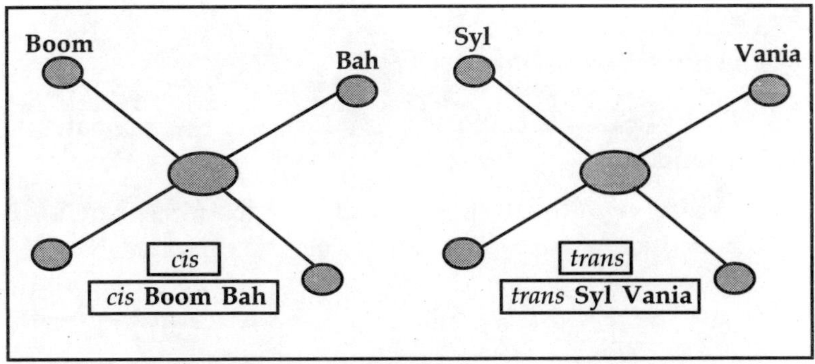

Fig. 14.1

Optical Isomers

It differ in the arrangement of four groups around a chiral carbon. These two isomers are differentiated as L and D.

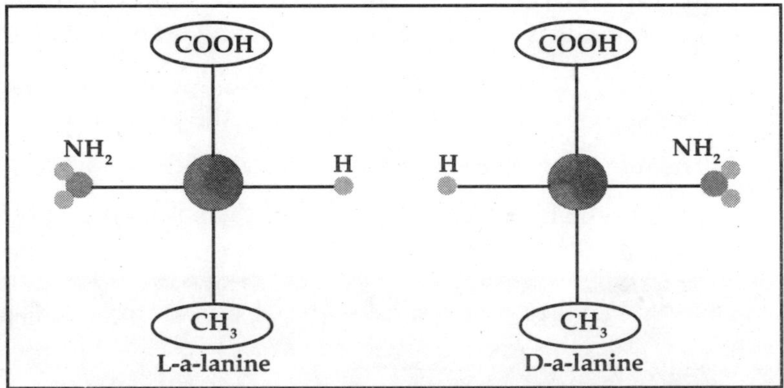

Fig. 14.2

Structural Isomers

Structural Isomers differentiate between the placement of two chlorine atoms around a hexagonal carbon ring. These three isomers are identified as *o, m,* and *p.* Once again we have given you a few clues to help your memory.

A pop-up nomenclature calculator is available for help when naming compounds and for practice problems.

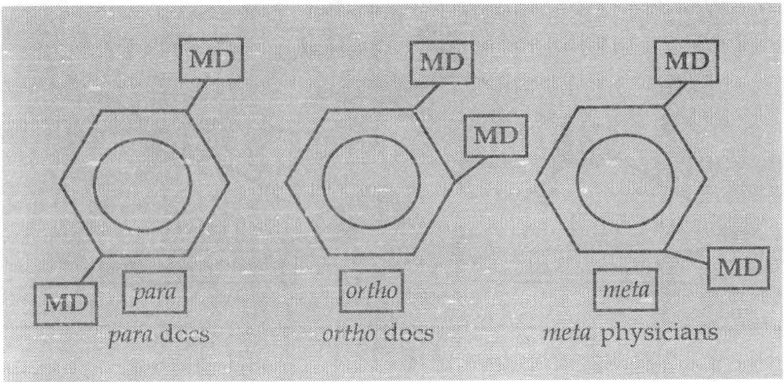

Fig 14.3

Naming Binary Molecular Compounds

Molecular compounds are formed from the covalent bonding between non-metallic elements. The nomenclature for these compounds is described in the following set of rules.

The more positive atom is written first (the atom which is the furthest to the left and to the bottom of the periodic table).

The more negative second atom has an "-ide" ending.

Each prefix indicates the number of each atom present in the compound.

Number of Atoms	Prefix
1.	mono
2.	di
3.	tri
4.	tetra
5.	penta
6.	hexa
7.	hepta
8.	octa
9.	nona
10.	deca

Examples:

CO_2 = carbon dioxide

P_4S_{10} = tetraphosphorus decasulfide

Naming Inorganic Acids

Binary acids (H plus a nonmetal element) are acids that dissociate into hydrogen atoms and anions in water. Acids that only release one hydrogen atom are known as *monoprotic.* Those acids that release more than one hydrogen atom are called *polyprotic*acids. When naming these binary acids, you merely add "hydro-" (denoting the presence of a hydrogen atom) to the beginning and "-ic acid" to the end of the anion name.

Examples:

HCl = hydrochloric acid

HBr = hydrobromic acid

Ternary acids (also called oxoacids, are formed by hydrogen plus another element plus oxygen) are based on the name of the anion. In this case, the *-ate,* and *-ite* suffixes for the anion are replaced with *-ic* and *-ous* respectively. The new anion name is then followed by the word "acid." The chart below depicts the changes in nomenclature.

Anion name	Acid name
hypo__ite	hypo__ous acid
__ite	__ous acid
__ate	__ic acid
per__ate	per__ic acid

Example:

ClO_4^- to $HClO_4$ => perchlorate to perchloric acid

ClO^- to HClO => hypochlorite to hypochlorous acid

Naming Compounds

A detailed treatise on naming organic compounds is beyond the scope of these materials, but some basics are presented. The wise chemistry student should consider memorizing the prefixes of the first ten organic compounds:

Number of Carbons	Prefix
1.	meth-
2.	eth-
3.	prop-
4.	but-
5.	pent-
6.	hex-
7.	hept-
8.	oct-
9.	non-
10.	dec-

There are four basic types of organic hydrocarbons, those chemicals with only carbon and hydrogen:

- Single bonds (alkane): suffix is "ane", formula C_nH2_{n+2}
- Double bonds (alkene): suffix is "ene", formula C_nH_{2n}
- Triple bonds (alkyne): suffix is "yne", formula CnH_{2n-2}

Cyclic compounds: use prefix "cyclo"

So, for example, an organic compound with the formula "C_6H_{14}" would be recognized as an alkane with six carbons, so its name is "hexane".

Examples:

N_2O_4 = dinitrogen tetraoxide

S_2F_{10} = disulfur decafluoride

Index

C

D

E